Coastal and Marine Environments - Physical Processes and Numerical Modelling

Edited by José Simão Antunes Do Carmo

Published in London, United Kingdom

IntechOpen

Supporting open minds since 2005

Coastal and Marine Environments - Physical Processes and Numerical Modelling
http://dx.doi.org/10.5772/intechopen.77408
Edited by José Simão Antunes Do Carmo

Contributors
Mona Fouad Kaiser, Walaa Awaad Ali, Maysara Khairy El Tahan, Laurence C. Breaker, Isabel Iglesias, Paulo Avilez-Valente, José Luís Pinho, Ana Bio, José Manuel Vieira, Luísa Bastos, Fernando Veloso-Gomes, Krishna Kishore Osuri, Nanda Kishore Reddy Busireddy, Kumar Ankur, Ljiljana R. Cander, Bruno Zolesi, José Simão Antunes Do Carmo

First published in London, United Kingdom, 2020 by IntechOpen
IntechOpen is the global imprint of INTECHOPEN LIMITED, registered in England and Wales, registration number: 11086078, 7th floor, 10 Lower Thames Street, London, EC3R 6AF, United Kingdom
Printed in Croatia

British Library Cataloguing-in-Publication Data
A catalogue record for this book is available from the British Library

Additional hard and PDF copies can be obtained from orders@intechopen.com

Coastal and Marine Environments - Physical Processes and Numerical Modelling
Edited by José Simão Antunes Do Carmo
p. cm.
Print ISBN 978-1-78984-359-0
Online ISBN 978-1-78984-360-6
eBook (PDF) ISBN 978-1-78984-732-1

For EU product safety concerns:
IN TECH d.o.o., Prolaz Marije Krucifikse Kozulić 3, 51000 Rijeka, Croatia, info@intechopen.com or visit our website at intechopen.com.

Meet the editor

José Simão Antunes do Carmo completed his Master's degree in Hydraulics and Water Resources in 1990 from the University of Lisbon and his PhD in Engineering Sciences in 1995 from the University of Coimbra, Portugal. He was director of several undergraduate and master courses in Civil Engineering and Environmental Engineering in the period 1995–2010. He was a scientific advisor of 24 Master's dissertations and two PhD theses in Hydraulics and Water Resources. He has been a member of the Ocean and Coastal Management Editorial Board since 2012. He has published two books, 63 papers in scientific journals, 14 book chapters, and more than 100 papers in international conferences. He has edited 10 books and a special issue of a scientific journal. His main areas of scientific research are hydrodynamics, morphodynamics, coastal management, river and coastal processes, climate change, natural hazards, risks, and vulnerabilities.

Contents

Preface XI

Section 1
Tools and Methodologies to Manage Coastal and Marine Environments 1

Chapter 1 3
Coastal Adaptation: Past Behaviors, Contemporary Management, and Future Options
by José Simão Antunes do Carmo

Chapter 2 23
Numerical Modeling Tools Applied to Estuarine and Coastal Hydrodynamics: A User Perspective
by Isabel Iglesias, Paulo Avilez-Valente, José Luís Pinho, Ana Bio, José Manuel Vieira, Luísa Bastos and Fernando Veloso-Gomes

Section 2
Coastal Dynamics and Ongoing Climate Processes in Coastal and Marine Environments 43

Chapter 3 45
Long-Term Changes in Sea Surface Temperature Off the Coast of Central California and Monterey Bay from 1920 to 2014: Are They Commensurate?
by Laurence C. Breaker

Chapter 4 75
Significance of Mesoscale Warm Core Eddy on Marine and Coastal Environment of the Bay of Bengal
by Nanda Kishore Reddy Busireddy, Kumar Ankur and Krishna Kishore Osuri

Section 3
Coastal Processes and HF Communications in Coastal and Marine Environments 89

Chapter 5 91
Modeling of Coastal Processes in the Mediterranean Sea: A Pilot Study on the Entrance of Suez Canal in Egypt
by Mona Fouad Kaiser, Walaa Awaad Ali and Maysara Khairy El Tahan

Chapter 6 **105**
Ionospheric Monitoring and Modeling Applicable to Coastal and Marine Environments
by Ljiljana R. Cander and Bruno Zolesi

Preface

It is well known that vulnerabilities and risks in coastal areas have been increasing, particularly since the middle of the last century, and a more marked increase is anticipated after the middle of the current century. It is also clear that human action has been the primary cause of the current imbalances, both directly (through local actions) and indirectly (through contributions to global warming and climate change). Adjustments in natural or human systems are thus necessary to respond to present-day or expected climate changes and their effects. The efforts made to reduce the causes and mitigate the effects of global climate change continue to be critical in coastal areas. Many adaptation strategies implemented in coastal areas remain inadequate or ineffective.

Sometimes for lack of financial support and others due to lack of scientific knowledge and technical preparation, the truth is that we continue to witness flooding, land loss, and human lives in many coastal areas around the world. However, much of the loss and damage that often occurs could have been prevented. Indeed, we may learn from the past that most adaptation measures have been reactive rather than proactive. It is generally understood, and the practice has been showing that, when planned, anticipatory adaptation will be more cost effective and efficient in the long term. It follows, therefore, that identifying and addressing needs and gaps in policies and planning will strengthen the adaptive capacity of regions and local communities.

On the other hand, as has been widely reported in the literature, the interactions between three main entities, land, human, and ocean, have been manifested positively and negatively in different contexts. Many studies have shown that the more human activities change, dominate, or even replace natural ecosystems, the more environmental vulnerability increases, and both the occurrence and scale of extreme events will be exacerbated. In the future, growth and concentration of populations in coastal cities and an aging population will contribute to increase further the vulnerability of cities to climate change.

To a large extent, direct and indirect consequences due to human activities are particularly serious for maintaining a sustainable balance, and hence the need to regulate human activities and reduce risk to achieve sustainable development of coastal zones. Although there is some perception of imbalances and vulnerabilities, there is still not a real awareness of the negative effects of human activities on such fundamental environments for the survival of humanity in the medium and long terms.

Due to seasonal factors and also to touristic reasons, the coastal communities of some coastal zones increase two or even three times in some periods of the year. Frequently, these communities rely on groundwater for their water supplies. In such circumstances, an increase in water pumping can have a marked effect on the position of the saltwater/freshwater interface. Lowering the freshwater head encourages the incursion of saltwater, which can rapidly render the supply undrinkable.

The removal or lowering of coastal dunes has a similar effect, as is the case for the reduction of the dunes' width. Dunes both act as recharge zones and support elevated water tables.

The use of coastal land for agriculture, building ports and harbors, industrial infrastructure development, and tourist urbanization are just some of a wide range of human activities affecting Europe's coasts. However, population growth and economic development are critical factors for change in coastal zones, which generate high pressure on ecosystems and natural resources due to increased use and proliferation of services. According to the European Environment Agency, between 1995 and 2025, the projected urbanization of the coastal zone on some coasts of the Mediterranean shows a built occupation increasing from 55% to 73% in Spain, 24% to 34% in France, and 38% to 45% in Italy.

The combined effects of land reclamation and its associated flood defenses, as well as protecting mobile habitats from erosion, have resulted in many kilometers of sea banks, seawalls, and groyne fields. The natural erosion of cliffs threatens individual buildings, towns, and villages, which are often protected by massive concrete seawalls and revetments as a result. The reduction in long-shore drift by the erection of these structures can lead to increases in erosion elsewhere, because the beaches are deprived of sediment. As long as the sea level continues to rise in response to global warming, increasing the height and resilience of hard coastal defenses will not ensure the protection of urban areas and territory for many years.

On the other hand, it should be noted that all interventions performed in coastal zones tend to be expensive, particularly when performed with a single goal (protection, for example); thus, it is crucial to address more clearly and with greater depth forms of social and economic intervention that are acceptable to stakeholders and local communities. However, coastal managers and policymakers should make effective and timely decisions on the use of appropriate adaptation measures for the immediate and longer terms. In other words, timely decisions should be taken on the measures to be implemented, with the objective of providing benefits immediately and also being effective enough to face future challenges.

It is therefore of utmost importance to bear in mind that reconciling the current activities in coastal zones with the maintenance of healthy ecosystems requires monitoring and the systematic evaluation and implementation of corrective measures. That is, coastal managers should be aware that simply implementing adaptation technology is not an endpoint; rather, adaptation is an ongoing process that requires constant prioritization of risks and opportunities, implementation of risk reduction measures, and review of their effectiveness.

Among the possible solutions to be implemented, it should be borne in mind that most adaptation measures can help in achieving multiple objectives and benefits. These measures can address current vulnerabilities and focus on increasing the capacity of ecosystems and communities to address the current environmental pressures and climate variability.

With all these issues in mind, it is obvious that an ideal approach to coastal zone planning and management should include the production of well-integrated action plans and emergency plans, as well as the monitoring and implementation of emergency warning systems. In this planning process, public participation is crucial. Coastal communities, non-governmental organizations, stakeholders, and investors should be involved in management and decision-making processes. In fact, all these actors should be considered part of the planning actions so that they are kept informed, motivated, and remain active during the various stages of defining the actions and measures to be implemented.

It is in this overall context that this book systematizes the concepts of contemporary coastal zone management and suggests possible structural and non-structural management tools for decision-making processes. Some successful adaptation measures and case studies on oceanic processes and coastal protection are also discussed.

It consists of six chapters, two of which address coastal management perspectives, adaptation measures, future accommodation options, tools, and methodologies to manage coastal and marine environments; two others focus on ocean dynamics and coastal processes, mainly due to changes in sea surface temperature and the effect of a mesoscale warm-core eddy on the marine and coastal environment of a bay; the latter two provide useful results and discussions on coastal processes that develop along the Suez Canal jetties using remote sensing and numerical modeling, and a review of the concepts and evolution of long-distance high-frequency (HF) communications in coastal and marine environments, with emphasis on HF radio communication during emergency situations and use of HF frequencies reflected by the ionosphere to detect objects at very long distances.

The selected chapters, which were identified to provide useful information and scientific knowledge, underwent a rigorous review process. They all contribute relevant information and useful content to scientists and other readers interested or concerned about the lack of adequate management actions and the installation of appropriate protection and adaptation measures or their ineffectiveness in containing coastal vulnerabilities and risks.

I believe this book comprises material of enough quality and quantity to make it a reference in the field of coastal and marine environments.

I would like to thank all collaborators who directly or indirectly helped set up this project, especially the guest referees and my Author Service Manager, Ms. Rozmari Marijan, for the opportunity they gave me to work together.

José Simão Antunes do Carmo
University of Coimbra,
Coimbra, Portugal

Section 1

Tools and Methodologies to Manage Coastal and Marine Environments

Chapter 1

Coastal Adaptation: Past Behaviors, Contemporary Management, and Future Options

José Simão Antunes do Carmo

Abstract

In the recent past, coastal public works solutions were generally designed as engineering problems. By that time, prior to the 1980s, the primary goal of coastal works projects was to maximize safety, taking into account only engineering knowledge and existing economic constraints. Today, concerns are no longer limited to safety; lifestyle and quality of life have become essential ingredients in building a successful coastal works project. Other aspects of the project are also important, such as environmental impact, attractiveness, and sustainability. These additionalcomplexities are further aggravated by other pieces of the puzzle that need to be integrated into the overall design, such as the non-engineering and non-science aspects. A synthesis of recent concerns regarding coastal public works projects has, in fact, become much more difficult for engineers to manage due to new assumptions of value, social acceptance, and sustainability of these projects. In this context, it is common knowledge that decision-making on a coastal issue should be based on criteria such as technical effectiveness, costs, benefits, implementation, and monitoring. This chapter addresses coastal issues using a dual perspective of meeting current needs and ensuring future sustainability. Contemporary adaptation measures and future accommodation options are also discussed.

Keywords: coastal issues, decision-making processes, public participation, coastal protection, adaptation strategies, future accommodation

1. Introduction

Coastal zones are interface regions between the mainland and the sea that are dominated by (1) processes that originate in the drainage basins of tributaries, (2) oceanographic and atmospheric processes, and (3) anthropogenic activities at different orders of magnitude and scales.

As recipient bodies, coastal areas not only receive the benefits of proper river basin management, but also suffer from the harm associated with or resulting from inefficient management processes. In particular, low water quality, sediment extraction, and sediment retention in structures implanted in the fluvial systems are the most evident factors that affect the use, natural resources, and activities that may occur in the coastal zones. The small amounts of sediment that flow into coastal areas are a matter of great concern to coastal managers, since this lack of sedimentation results in the disappearance or devaluation of beaches and the weakening of natural protection systems.

To take into account these concerns and other concerns resulting from ongoing and future impacts from climate change, managers must approach coastal zones management different than they did in the past. Inadequate management methodologies are still common practices today in many countries that have fewer resources or less concern about global climate change.

We may learn from the past that most defensive measures have been more reactive than proactive. In addition, the intervention procedures in the coastal zones, whether for the purposes of valuation or implementation of protection measures, should be improved.

Until recently, much of the coastal engineering projects were almost exclusively based on experience. As discussed in Kamphuis [1], it was in this context that many of the coastal defense works were carried out and resulted in kilometers of breakwaters, sea walls, and groyne fields for defenses against flooding and urban fronts, recovery of degraded areas, and protection of heritage of great historical and cultural value.

It is common knowledge that hard engineering structures can be effective when properly designed and installed. However, because they are continuously subjected to events that in many cases exceed their design capacity, such structures require adequate and costly maintenance. On the other hand, it is also known that these structures tend to reduce erosion and the risk of flooding in one location, but increasing the risk in another.

Meanwhile, with the evolution of the numerical methods and the computational means, it has become possible to use increasingly sophisticated mathematical models to solve complex coastal engineering problems. The use of ever more powerful, less restrictive and more user-friendly computational models, along with physical modeling, is today a common practice in coastal engineering.

On the other hand, it is of utmost importance to involve interdisciplinary groups that cover different perspectives (such as policy makers, civil and environmental engineers, geologists, biologists, economists, sociologists, lawyers, etc.). These interdisciplinary groups should include local communities and stakeholders at different stages of project development, including design, building, and monitoring.

Indeed, public participation, including stakeholders, is considered as a key principle when planning and implementing conservation projects. The same view is shared by Hedelin et al. [2] and Barceló [3], who clarifies in an Elsevier editorial note *when people are ignored and conservation measures are put in, we see opposition, conflict and often failure. These problems require the best available evidence, and that includes having both natural and social scientists at the table*.

In this context, the guidelines on coastal defenses have changed. Contrary to the coastal defenses built in the past, contemporary adaptation measures include: artificial sand nourishments, possibly with additional sand support measures; building and rehabilitation of dunes; creation and restoration of wetlands; reinforcement or creation of submerged longitudinal bars; submerged breakwaters made of geotextile tubes; buffer zones; and land use restriction and zoning.

Other measures, which are less common, but may become essential in the near future, include: building on pilings; adaptation of drainage systems; building emergency flood shelters; and tidal houses, houseboats, and floating houses.

Reconciling current activities in the coastal zones with the maintenance of healthy ecosystems requires monitoring, systematic evaluation, and implementation of corrective measures. Indeed, it is generally understood and practice has shown that when planned, preventative adaptation will be more cost-effective and efficient in the long run than retroactive measures. Therefore, identifying and addressing needs and gaps in policies and planning will strengthen the adaptive capacity of regions and local communities.

In economic terms, coastal zones' natural wealth and the wide diversity of activities taking place make these regions the main source of revenue for many countries. In fact, coastal zones are currently (1) important areas of food production through agriculture, fishing, and aquaculture, (2) the main tourist destinations on all continents, (3) significant sources of mineral resources, including oil and natural gas, (4) foci of industrial development and transport, and (5) abundant reservoirs of biodiversity and ecosystems, on which the functioning of the planet depends.

However, some decreases in attractiveness or decreases in the demand rate have been observed. On one hand, these decreases will contribute to the maintenance and sustainability of coastal ecosystems, but such behavior can also result in socially unsustainable conditions.

In fact, unfavorable circumstances in coastal zones include: (1) large concentrations of people and services in sensitive or risk areas, leading to artificialization of certain stretches; (2) insufficient or inefficient services (housing, security, health, catering, banks, leisure, bathing, etc.) to meet more needs during high-demand seasons; (3) scarcity (in quantity and quality) of water resources in seasons of increased demand and consumption; (4) great specialization of some economic activities, directed to very specific users and in very restricted periods of the year; (5) a poorly designed, non-existent, or very permissive arrangement of spaces with unbridled/abusive and uncontrolled occupations; (6) greater speculation, with uncontrolled costs and often incompatible with the quality of the services provided; (7) numerous situations of stress/confusion incompatible with rest (e.g., physical and mental recovery) that are sought; and (8) of least concern, the interests and well-being of the residents throughout different times of the year.

It is indispensable to ensure the availability of the amount and quality of water resources needed. However, excessive consumption of water beyond sustainable availability can lead to irreversible degradation. Particularly worrying is the contamination of groundwater due to excess of water consumption from coastal aquifers, leading to exaggerated lowering of groundwater levels and the salinization of these waters. For example, **Figure 1** shows stains of the wells and boreholes that exist in the Algarve coastal zone (southern Portugal) and 17 aquifers (M1, M2,..., M17) with regional expression [4].

The impact of saline intrusion on freshwater aquifers and the penetration of salt water in the estuaries will be substantially aggravated by global warming and the consequent rise in mean sea level. Due to excessive water consumption, coastal aquifers contamination by saltwater intrusion is already a reality in some southern European countries. Portugal is an example, particularly in the southern coastal region of the Algarve (**Figure 1**).

To cope with water scarcity in the months of greater tourist influx, in which the population reaches at least three times the resident population, two dams were built in the mid-twentieth century (Arade and Bravura). Subsequently, a third dam was built in the 1980s (Beliche), which was followed by the building of two more dams in the 1990s (Funcho and Odeleite), and more recently a sixth (Odelouca).

The heights of these dams vary between 41 and over 90 m and with maximum storage capacities between 2.7×10^7 and 13.5×10^7 m^3 [4]. However, these dams also suffer the effects of water shortages in the hottest months of the year, which coincide with periods of higher consumption.

In addition, high pollutant loads discharged directly into the sea without any treatment or with an inadequate level of treatment contaminate or pollute the sea, thereby excluding these waters from bathing uses and natural resources conservation, with considerable environmental impacts. Often, changes in environmental factors give rise to qualitative changes in established ecosystems (e.g., specific composition, biodiversity, etc.).

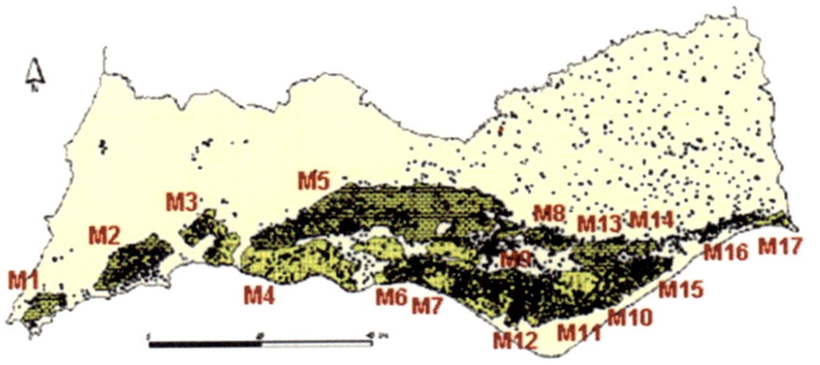

Figure 1.
Well and borehole stains (blue points) and 17 aquifer systems (Mx) that exist with regional expression in Algarve, Portugal (Adapted from [4]).

2. Traditional and contemporary decision-making procedures in coastal management processes

From the procedural point of view, the implementation of works in the coastal zone followed a very simple procedure in a not too distant past. All interventions were focused on the project, which was entirely managed by an engineer. This engineer was responsible for everything and only had contacts with the entity responsible for coastal works. The contribution of experts from disciplinary areas with different perspectives did not exist or was very limited. In fact, until the 1980s, the construction of works in coastal zones essentially followed the schematic diagram shown in **Figure 2**.

At that time, the coastal public works were contracted and overseen by the project owner, usually a government entity or a construction/business company, who alone was responsible for project decisions, coastal work implementation, and monitoring of its behavior [1]. Coastal science at that time was essentially physical (hydrodynamics—waves, currents, tides, etc.).

In Portugal, some coastal public works projects that were executed in this context are noteworthy. Until the 1980s, the entire process was carried out in accordance with the procedures shown in **Figure 2**. The interventions were usually based

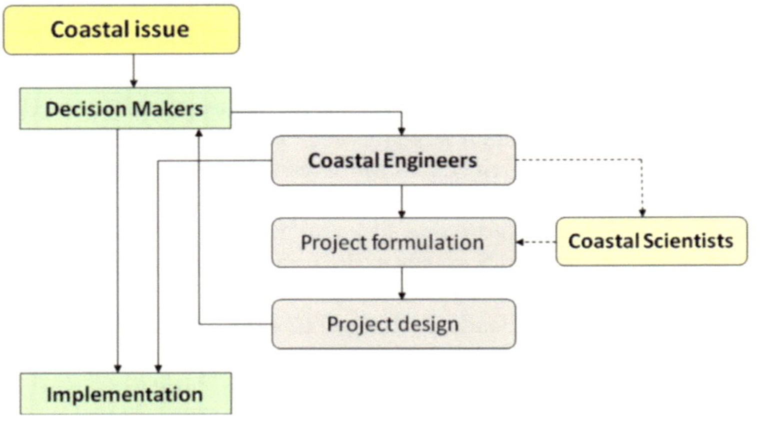

Figure 2.
Traditional decision-making process in coastal zone issues (Adapted from [1]).

on a structural project and lacked impact studies, environmental concerns, public consultation, and intervention/incorporation of the input from local communities and stakeholders.

Figure 3 shows some of the interventions undertaken on the west coast of Portugal in the 1970s and 1980s and with subsequent structural reinforcement. Currently, the urban centers shown in this figure are residential areas at high vulnerability and risk, only maintained at the expense of hard engineering projects.

It should be noted that the option for hard engineering projects was motivated by the general erosion trend of the Portuguese Atlantic coast and, in particular, by the defense of urban fronts. However, in general, these hard structures have had negative consequences along the Portuguese coast, as they have increased coastal erosion in remote areas by impeding the normal circulation of sediments. These projects also led to negative consequences locally as they led to urban expansions often occupying marginal areas as a result of false safety sensations provided by such structures (**Figure 3**).

In the 1980s, the development of more in-depth theoretical knowledge and the evolution of numerical methods and computer hardware occurred. These developments allowed engineers to develop and apply numerical models capable of describing the physical processes with greater accuracy.

It was in this context that the field of hydroinformatics emerged in the 1990s, which is a new scientific branch that links informatics tools, hydraulics, environmental concepts, and models with the overall objective of solving environmental problems in coastal waters. Examples of hydroinformatic environments created for this purpose are the modular structures described in Pinho et al. [6] and Deltares [7–9] (DELFT3D), which are capable of simulating hydrodynamic, morphodynamic, and water quality processes.

These modular structures and other highly complex two- and three-dimensional computational structures are now commonly used to solve real-world problems, particularly in coastal areas. Among the most common are: MIKE21 (https://www.mikepoweredbydhi.com/products/mike-21), POM (http://www.ccpo.odu.edu/POMWEB/), ADCIRC (http://adcirc.org/), TELEMAC3D (http://www.opentelemac.org/), DELFT3D (https://oss.deltares.nl/web/delft3d), and CCHE3D (https://www.ncche.olemiss.edu/cche3d).

Meanwhile, as addressed in [1], concerns in coastal areas are no longer limited to safety issues: lifestyle and quality of life have become essential ingredients. Other aspects of the project have also become important, such as environmental impact, esthetics, and sustainability. The additional complexities were further aggravated by other aspects that needed to be integrated into the overall design, such as

Figure 3.
Urban seafronts protected by sea walls and groins. These residential areas are located at the western Portuguese coast (Mira on left and Esmoriz-Cortegaça on right) and are only maintained at the expense of hard engineering projects (Courtesy of Lopes [5]).

non-engineering and non-science concerns. Examples of these concerns include socio-economic aspects and quality of life, involving leisure, tourism, sporting practices, fishing industries, water quality, etc.

These emerging sociological realities needed to be addressed, as well as the voices of actors and interest groups that would like their input incorporated into the project design. In fact, a synthesis of these recent concerns has become much more difficult to manage since this new reality is based on assumptions of value, social acceptance, and sustainability. The interrelationships presented in **Figure 4** show the current complexity inherent to the contemporary management process of the coastal zone.

The organizational chart shown in **Figure 4** shows that to integrate all social and technical requirements and to facilitate an optimum solution, coastal managers must organize and maintain clear communication between the various actors.

As is clear from preceding discussion, two distinct realities stand out from the traditional and contemporary approaches to the coastal planning and management issues. On one hand, the need to utilize more scientific and technological knowledge into addressing coastal issues is recognized. Therefore, specialists from different disciplines enrich the structural component of engineering design. On the other hand, the need to involve public agents, entities, interest groups, and local communities is recognized in order to ensure the necessary support and social component of the structural component.

These concepts synthesize the current manner of addressing with the coastal issues. The interactivity between both the physical-environmental and socio-economic systems is at the interface of the well-known concept of Integrated Coastal Zone Management, which requires the integration of different disciplinary expertise from local, societal, and practical knowledge within coastal planning and decision-making processes.

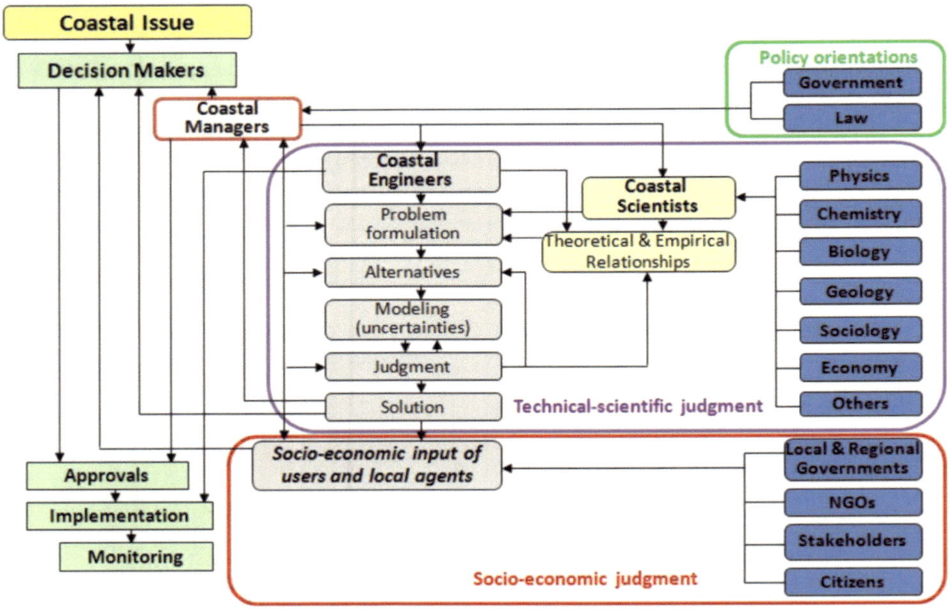

Figure 4.
Contemporary decision-making process in coastal zone issues (Adapted from [1]).

Figure 5 shows how these systems complement each other for the success of an intervention located in the coastal zone. The structural component (or resistant structure) suffers the physical effects of the processes and is fundamentally of the technical-scientific and environmental domain (PES). The socio-economic component (SES) forms the base of support of the resistant structure and is the domain of institutions, stakeholders, and citizens.

The need to involve researchers from different disciplinary areas in the conceptualization and development phases of an intervention program in the coastal zone has been recognized in several coastal interventions carried out on the Portuguese coast. The same has occurred with the need to involve public institutions, local communities, users, and citizens in general in the process of socio-economic consultation.

In this context, the intervention carried out at Costa da Caparica [11], which is an extensive well-attended bathing area, located near Lisbon, Portugal, shown in **Figure 6**, was particularly relevant.

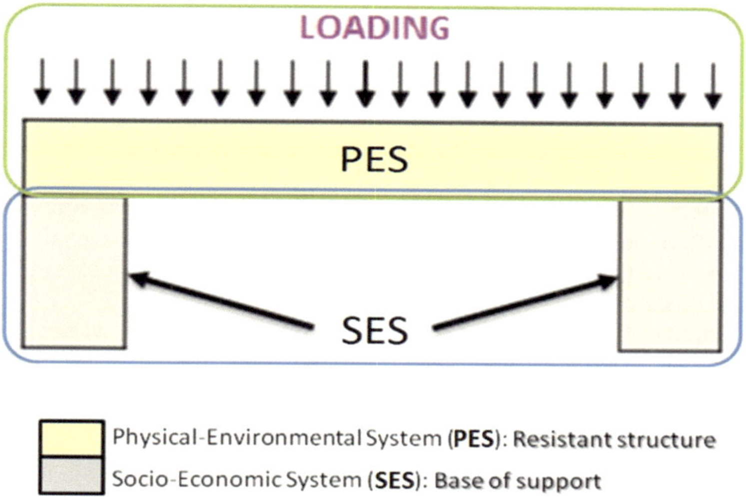

Figure 5.
Typical coastal system composed of a resistant structure and its base of support (Adapted from [10]).

Figure 6.
Views of Costa da Caparica beach, close to Lisbon, Portugal, deprived of sand in 2006 (Courtesy of Alveirinho Dias), and the beach and its marginal strip after the last major intervention in 2014, on the left and right, respectively (Adapted from [12]).

3. Levels of operation and public participation in coastal issues decisions

As is clear from the foregoing considerations, the inclusion of different disciplinary groups is of the utmost importance for the success of any intervention program in the coastal zone. In fact, contemporary assessment processes are based on vulnerability indexes or coastal sensitivity, which are functions of several variables or physical parameters that require a diversified knowledge base that has a depth that goes beyond the pure domain of engineering.

In order to be successful, managers must consider not only physical processes and economic interests, but also the opinions and participation of citizens, stakeholders, and local communities in any planning process, conservation project, and coastal development. Public participation is paramount to ensure the development and sustainability of any coastal zone. The use of management strategies that address and consider public perception of the environmental risks, erosion effects, cyclones, tidal surges, and floods is appropriate.

However, the inclusion of entities and people to have a passive attitude is not enough. Managers must also reflect on the level of participation of the agents involved. As is reported in Guimarães et al. [13], seven levels of participation can be defined: passive, in which participants are informed of what will happen; informatory, which provides answers to participants' questions; consultation, where the participants are consulted and their perspectives are heard; incentives, in which people participate for incentives; functional, in which groups are formed that aim to achieve defined objectives; interactive, in which people participate in joint analyses to define actions; and finally, mobilizing participation, in which people participate by taking initiatives independently of external institutions. These levels of participation correspond to different levels of interaction and can be considered distinct stages of the decision-making process.

This brief analysis highlights the need to involve many actors when implementing procedures in order to produce well-accepted and sufficiently credible decision-making vulnerability and risk assessment projects in the coastal zone. However, these procedures will only succeed with as much consensus as possible, which should be useful for integrated planning and management of the coastal zone; thus, these procedures serve to establish priorities for intervention.

Truly shared management corresponds to the levels of involvement with a high degree of interaction, which encourages various types of participation, simultaneously functional, interactive, and mobilizing. For this reason, it is essential to establish trusting relationships that must be supported with dialog and a discourse among the different groups involved in the decision-making process. This relational procedure is primarily based on three dimensions: integrate, interact, and inform, as is schematically shown in **Figure 7**.

The process begins by identifying the intervention needs, which are accomplished by defining the type and design of the project. The intervention needs are noted during the implementation period of the coastal work and continue with the monitoring of the structure and surrounding space.

In all of the stages, various stakeholders should be involved and heard in the decision-making process. Decision makers should consider stakeholders an integral part of the plan to remain informed, motivated, and active during the various phases of definition, implementation, and monitoring of the proposed project.

The creation and maintenance of a healthy, multi-functional system requires strong collaboration between broadly skilled technicians, public and private entities, local authorities, residents, non-governmental organizations (NGOs), stakeholders, and citizens (**Figure 4**).

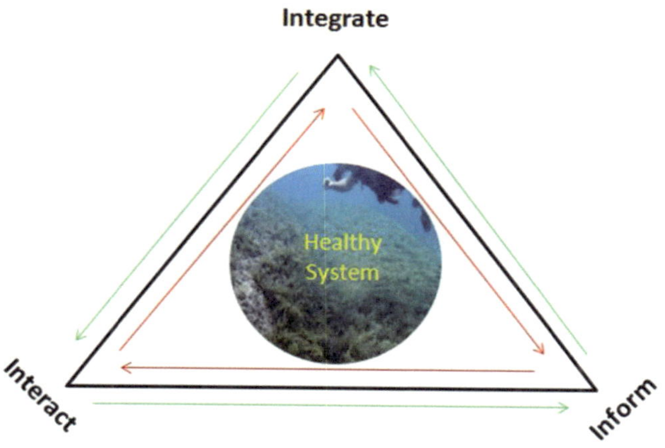

Figure 7.
Three key-dimensions for the success of the Integrated Coastal Zone Management.

In the process of developing a coastal intervention, setting commitments that result in the integration and involvement of all potential stakeholders is essential for a manager to be successful in the decision-making process. With this same goal, sensitizing the agents involved to the opportunities, potential failures, and lower individual gains in favor of the collective benefit is also important.

A permanent interaction is also essential to establish a framework for reciprocal cooperation between all parties. This interaction will provide clear and transparent information on the possible options, in order to ensure that all parties equally accept, share, and assume the expected benefits, costs, and risks involved with the project.

4. Prospects for future accommodation in coastal areas

Contemporary coastal managers are concerned with the preservation and enjoyment of coastal zones. Proper use, assessment, and monitoring of natural resources are important goals for preservation. However, to what extent will the current tourist densification in coastal zones be compatible with the sustainability of coastal zones? Will the complementary effects of global climate change lead to significant changes in the current demand for the coastal zones?

In fact, the following facts are true: (1) two-thirds of the world's megacities are located on the coast, and more than half of the population of the 22 European Member States (that have a coastline) live less than 50 km from the sea [14] and (2) these narrow coastal strips correspond to only approximately 10% of the living space on Earth. According to Berger [15], one billion people could live along the coasts, at or below 10-m elevations, by the year 2060.

Population growth and economic development are critical factors for change in coastal zones, which generate high pressure on ecosystems and natural resources due to increased use and proliferation of services. According to European Environment Agency [16], between 1995 and 2025, the projected urbanization of the coastal zone on some coasts of the Mediterranean shows a built occupation increasing from 55 to 73% in Spain, 24 to 34% in France, and 38 to 45% in Italy.

Failure to reverse this situation will inevitably lead to ecosystem collapse, which will occur more or less rapidly depending on the evolution/intensification of the following factors: (1) lack of pleasant spaces, leading to unbridled/abusive and uncontrolled occupations; (2) depletion of water resources (in quantity and quality); (3) inefficiency of services (security, health, catering, leisure, socializing, etc.); (4) increased speculation, with unbearable costs, and (5) aggravation of conflicts and insecurity.

In the meantime, maintaining coastal zone esthetics and sustainability in areas of increasing risk will only be possible through regular interventions, in accordance with the guidelines expressed in the previous points. Long-lasting sustainability can be achieved by implementing interventions that effectively reduce wave energy prior to reaching the coastline.

In fact, it is possible to guide the waves propagation by acting on the bathymetry in areas of the continental shelf, where the waves propagate in intermediate- and shallow-water conditions, forcing waves to rotate (refraction effect) and to break in positions away from the coastline; thus, preventing all of the wave energy from being discharged on the beach and/or in other natural defense systems, such as dunes. To preserve coastal dunes and stabilize the coastal foundation properly, one strategy recommends the use of artificial nourishments, possibly complemented by additional protections to prevent sand losses, such as longitudinal detached submerged breakwaters made of geotextile material. The goal of this strategy is to protect the coast in an environmentally friendly and esthetically pleasing manner. Examples of effective actions that protect the coast in this manner are presented in Oh and Shin [17] and Taal et al. [18].

It is important to note that coastal zone defense programs should adopt protection principles based on preventive actions of wave behavior, that is, interventions must be carried out from the ocean to the coast.

On this issue, the Leirosa case study has been an example of learning. This sand dune system has been the scene of three major dune rehabilitation interventions in the past 15 years. The first involved the reconstruction of the sand dunes followed by re-vegetation [19]; see **Figure 8**.

The second intervention consisted of installing geotextile containers filled with sand [20, 21], and the last one was implemented to stabilize the existing geotextile-reinforced sand dune system in the area where some encapsulated sand layers, mainly the bottom three, had partially opened up [22, 23]; see **Figure 9**.

This protection strategy allowed this stretch of the coast to remain more or less stable until 2014. However, the Hercules and Stephanie storms that struck the Portuguese coast in 2014 caused deep damages in this dune system, thus increasing the weaknesses that existed.

These events have taught us that the measures taken to protect the beach and dune system of Leirosa would not be sufficient, so we put forward a proposal to install a multi-functional submerged structure with characteristics to (1) protect

Figure 8.
Rehabilitated stretch of the Leirosa sand dunes system, from May to September 2000 (Adapted from [12]).

Figure 9.
General aspect of some bottom geotextile layers after the high tides late in the winter of 2006, in April (Adapted from [12]).

the coastal zone, dissipating energy of waves, (2) create a calmer sea on the lee side of the structure, and (3) increase the surfing possibilities in the Leirosa area of Portugal. In addition, it should be noted that coastal defense structures incorporating multi-functionalities are, in general, well-accepted by stakeholders [24]. Still according to Evans et al. [24], stakeholders recognize that the benefits to society provided by such structures could attract public and private funding.

The proposed structure for Leirosa was tested for two reef geometries with different reef angles (45 and 66°) [25, 26]. Wave data were obtained from a DATAWELL directional wave buoy located about 5 km off the coastline, position 40° 03′ 22″ N 8° 57′ 22″ W, at 25 m water depth. Average significant heights (HS) and wave peak directions (TZ) for 3-h intervals were recorded over a period of 9 months. Values of the peak period and wave direction distributions for different wave heights are shown in **Figure 10**.

Since these data series are very short, the records shown in **Figure 10** were compared to longer series (12 years) of average wave heights and peak periods obtained at a station located in Figueira da Foz, about 15 km to the north of Leirosa. These data were transposed to the wave buoy installed near Leirosa using the coastal wave model SWAN. Comparisons of frequency histograms for significant wave heights and peak periods obtained by both methods are shown in **Figure 11**.

Taking into account the results shown in **Figure 11**, where pronounced differences are noted especially in the wave periods, two design wave conditions were tested: typical storm conditions on the Portuguese west coast (wave height H = 4:0 m, period T = 15 s), and common conditions on this coast (H = 1:5 m, T = 9 s). Accordingly, the four scenarios C1–C4 shown in **Table 1** were simulated.

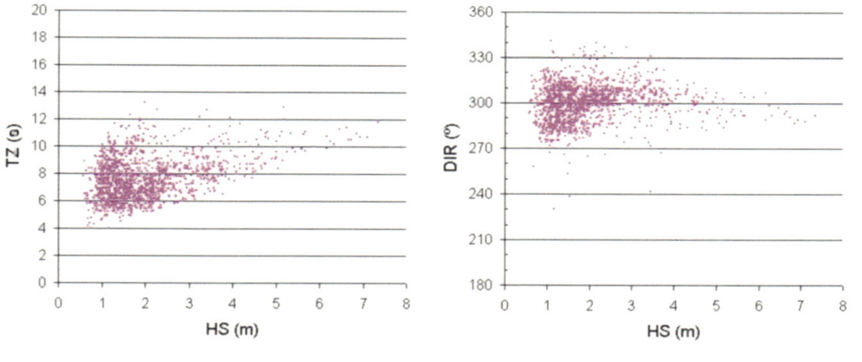

Figure 10.
Local wave data obtained from a DATAWELL directional wave buoy, at 25 m water depth: TZ-HS and DIR-HS relationships, on the left and right, respectively.

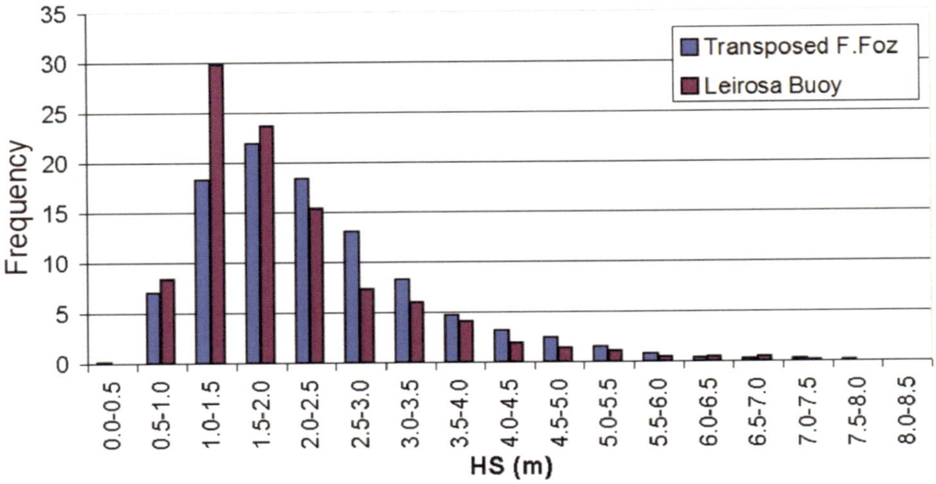

a)

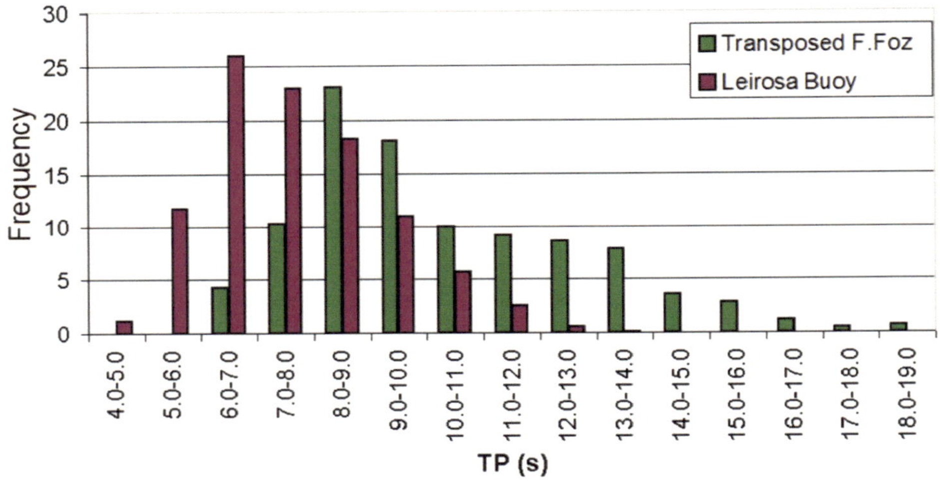

b)

Figure 11.
Frequency histograms for significant wave heights and peak periods. (a) Significant height, HS. (b) Peek period, TP.

In order to propagate the incident wave, a Boussinesq type model, COULWAVE [27] was used. **Figure 12** shows numerical results of wave heights and the wave breaking line for the common wave conditions and a reef angle = 45° (C1 test, **Table 1**).

Figure 12 shows the good performance of an artificial submerged reef taking advantage of the wave refraction effect. This structure also causes wave to break in the water mass (away from the coastline). Therefore, it is possible not only to increase the width of the beach (new position of the breaker line), but also take advantage of the generated wave characteristics (shoaling effects) for sports practices. This structure can be designed to meet current conditions and, if necessary, be further strengthened to take into account possible changes in coastal dynamics.

At a later stage, earlier actions will no longer be effective, possibly as the result of ongoing climate change. This change may lead to an increased demand for accommodation alternatives in coastal areas. In areas of greater scarcity and risk,

Scenario	Reef angle (°)	H (m)	T (s)	Points per wavelength	Grid size (m)	Time step (s)
C1	45	1.5	9	43	2.14	0.09288
C2	45	4.0	13	60	2.77	0.11999
C3	66	1.5	9	43	2.14	0.09288
C4	66	4.0	13	70	2.37	0.10285

Table 1.
Reef, wave, and mesh characteristics used for the four simulated scenarios [26].

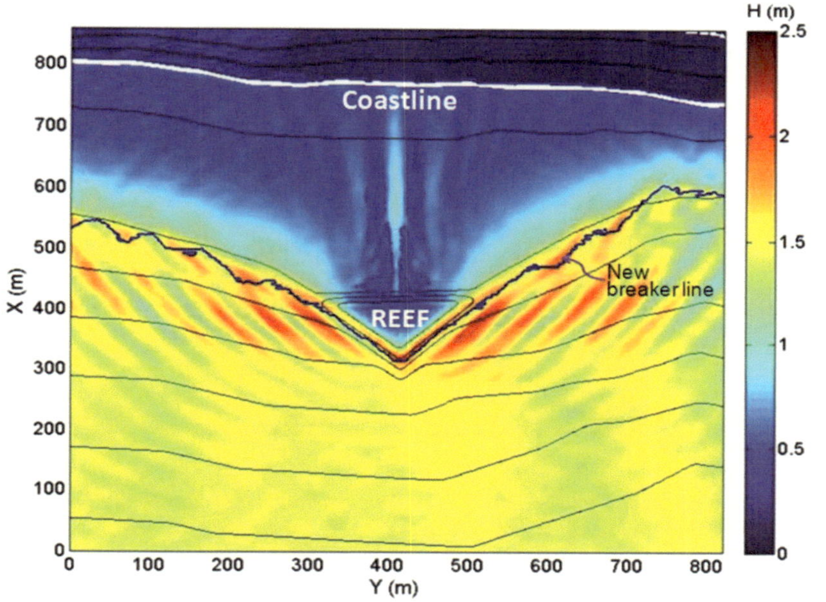

Figure 12.
Wave heights and wave breaking line around the reef area (reef angle = 45°; wave height H = 1.5 m, and period T = 9.0 s, as C1 test, **Table 1***) (Adapted from [26]).*

countries such as the USA, the UK, India, Indonesia, Philippines, Thailand, and many others already have elevated constructions on pilings.

In essence, it is the resumption of very old defense principles (e.g., stilts), although with potentially different motivations; however, these same flood defense purposes constructions are not new. As shown in **Figure 13**, the coastal areas of countries with extensive lowlands, such as Portugal, have constructions that were already adopted as flood defense options in the second half of the nineteenth century.

With the same goal of flood defense, demand and housing construction has grown significantly in several coastal regions of the world. For example, many houses are built on pilings along the west coast of the USA. **Figure 14** shows a set of houses of this type built on Malibu beach, California, USA, with waves crashing underneath the houses. **Figure 15** shows two luxurious waterfront homes built on pilings over the water, also in the USA coast.

More examples, among many others, are shown in https://www.homestratosphere.com/houses-built-on-stilts/. However, these structures require some care because, as noted in Park et al. [29], the structure elevation is a critical variable

Figure 13.
Flood-proof piling houses built on the coast of Tocha-Mira, Portugal, in the mid-nineteenth century (Adapted from [28]).

Figure 14.
Houses on Malibu beach built on pilings with waves crashing underneath the houses (https://www.homestratosphere.com/houses-built-on-stilts/).

Figure 15.
Two luxurious waterfront homes built on pilings over the water in the USA (https://www.homestratosphere.com/houses-built-on-stilts/).

affecting property damage and loss; they show that vertical or horizontal force caused by a given set of wave conditions may increase or decrease depending on the structure's elevation above the water level. Guidelines on how to design and build safer and less vulnerable housing to reduce the risk of life and property in coastal areas can be found in Coulbourne et al. [30].

More recently, a significant increase of floating houses and houseboats has been noted. Facilities of this kind are multiplying throughout the world, including

Figure 16.
Types of floating houses, as future widespread adaptation measure to address coastal hazards and climate change (Google, unknown author and date).

the sets of houseboats at De Omval, Amsterdam, on the river Amstel (just downstream from De Omval), the 60 houseboats situated on the north side of the River Thames, on Battersea Reach, and the harbor at Bembridge, with approximately 25 houseboats [12].

An increasing demand for this type of housing and a reduction of current living conditions in coastal areas are foreseeable, especially from the middle of the present century. As a consequence of global climate change, it is also foreseeable that an increase of integrated neighborhoods (including houseboats) in the coastal cities may occur. Alternatively, independent urban poles exclusively composed of houseboats, tidal houses, and floating houses like those shown in **Figure 16** may increase.

5. Conclusions

Coastal areas are now places of attraction for large masses. Coastal managers have a growing interest in implementing effective management that meets current needs, while also taking into account the need to sustainably manage natural resources for future generations. To satisfy these requirements, alternative solutions that safeguard the sustainability of the environment, tourism social resources, and services in the coastal zones will have to be considered.

Contemporary coastal management relies on the basis of integration and accountability of all stakeholders, including local communities, investors, technicians, specialists (from different disciplinary areas), and managers in the processes of conceptualization, decision-making, implementation, and monitoring of any intervention program in the coastal zone. Coastal management should aim to sensitize all stakeholders to the intervention needs, hazards, and inherent risks. Stakeholders and managers should discuss collegially the possible solutions and corresponding costs and participate in decision-making processes, accepting in this way a potential failure.

The phases of implementation and monitoring should be shared in such a way that everyone is proud of the success of the intervention or be motivated to accept and correct failure. However, if multiple interest groups are involved and these groups have conflicting interests, complexities can manifest themselves even in the acceptance and approval of processes, often resulting in conflicting voices [31]. These conflicts may be an additional problem, which may lead to only some of the stakeholders supporting the process.

Several factors will contribute to the need for adapting management procedures: global warming, consequent rise in mean sea levels, increased frequency, and intensification of storms, especially beginning in the middle of the present century [32]. These impacts from global warming will require other forms of accommodation in coastal areas.

At some point, it will no longer be possible to maintain the effectiveness of the protection measures in the high-risk areas. The effort to continue living in vulnerable areas, possibly even with high loss rates, will remain for some time, but there will come a time when the risk will no longer be acceptable and much of the effort will focus on retreating from these areas. At that time, local communities will eventually accept that retreat to a safer place is necessary. The question of "How long will the location where we retreated be a safer place?" will always remain. The costs involved in maintaining the effectiveness of protection measures in high-risk areas may also be an additional problem.

Measures of protection and accommodation will be adapted to the circumstances. The need to model the seabed on the continental shelf (at depths of 5–10 m), by redirecting the propagation of waves and forcing them to break onto the water mass (away from the coastline) is increasingly recognized. Consequently, by preventing waves from discharging much of their energy onto beaches and dune systems, natural protection systems will be less exposed.

Possible solutions to erosion include artificial sand nourishments with installation of submerged coastal control structures, such as submerged longitudinal bars, using sand-filled geotextile tubes as sand containment supports, as in Oh and Shin [17]. Another possible solution consists in the installation of submerged multi-functional artificial reefs equally with the use of geotextiles in tubular form, of which is example the Narrowneck reef installed in the Gold Coast, Australia, in response to the increasing occurrences of beach erosion [33].

Accommodation measures that use aquatic environments have long been a reality. More recently, the number tidal houses, houseboats, and floating houses are increasing. A wide variety of houseboat options are available, both in terms of features and dimensions.

The contemporary reality still allows resistance to adverse conditions with relatively soft adaptation solutions; however, most forecasts point to significant changes within a few decades [32]. High concentrations of population and services in coastal areas, increasing difficulties in finding safe and pleasant spaces and the expected flooding of many lowlands as a result of global climate change are favorable conditions for the search and installation of accommodation alternatives.

As endnotes, it is recommended the involvement of people (residents, citizens, stakeholders, and others), technical support and government in the actions to be developed under the following guidelines:

- Multi-functional flood defense infrastructures can be developed and should be implemented for the benefit of local people and businesses.

- Barriers are not enough as flood defenses—existing and new houses can become more resilient and flood resistant, and should be part of an integrated community or a city's entire strategy.

- The planning systems should encourage integrated solutions and innovative long-term local strategies.

DOI: http://dx.doi.org/10.5772/intechopen.88123

Author details

José Simão Antunes do Carmo
Department of Civil Engineering, University of Coimbra, Coimbra, Portugal

*Address all correspondence to: jsacarmo@de.uc.pt

References

[1] Kamphuis JW. Costal science, engineering and management. In: Proceedings of the Canadian Coastal Conference 2005; Dartmouth, Nova Scotia, Canada; 2005. 9p

[2] Hedelin B, Evers M, Alkan-Olsson J, Jonsson A. Participatory modelling for sustainable development: Key issues derived from five cases of natural resource and disaster risk management. Environmental Science and Policy. 2017;**76**:185-196. DOI: 10.1016/j.envsci.2017.07.001

[3] Barceló D. Researching New Ways to Connect People to Nature: A Special Collection for #WorldEnvironmentDay. Editor's Note. 2017. https://www.elsevier.com/connect/researching-new-ways-to-connect-people-to-nature-a-special-collection-for-worldenvironmentday

[4] Nunes L, Monteiro JP, Cunha MC, Vieira J, Lucas H, Ribeiro L. The water crisis in southern Portugal: how did we get there and how should we solve it. WIT Transactions on Ecology and the Environment. 2006;**99**:435-444. DOI: 10.2495/RAV060431

[5] Lopes AM. Information systems applied to the integration of knowledge in coastal management of the Central Region. In: Colloquium on Coastal Management: Vulnerabilities and Risks in the Central Region; Portugal: University of Aveiro, November 26, 2010

[6] Pinho JLS, Vieira JMP, Antunes do Carmo JS. Hydroinformatic environment for coastal waters hydrodynamics and water quality modelling. Journal of Advances in Engineering Software. 2004;**35**:205-222. DOI: 10.1016/j.advengsoft.2004.01.001

[7] Deltares. User Manual Delft3D-FLOW. Hydro-Morphodynamics, Version: 3.15.34158. The Netherlands: Deltares; 2014. https://oss.deltares.nl/documents/183920/185723/Delft3D-FLOW_User_Manual.pdf

[8] Deltares. User Manual Delft3D-WAVE. Hydro-Morphodynamics. Simulation of Short-Crested Waves with SWAN. Version: 3.05. The Netherlands: Deltares; 2019. https://content.oss.deltares.nl/delft3d/manuals/Delft3D-WAVE_User_Manual.pdf

[9] Deltares. User Manual D-Water Quality. Water quality and Aquatic Ecology. Version: 5.06. The Netherlands: Deltares; 2019. https://content.oss.deltares.nl/delft3d/manuals/D-Water_Quality_PLCT_User_Manual.pdf

[10] Kamphuis JW. Coastal engineering education and coastal models. Coastal Engineering Proceedings. 2012;**33**:30. DOI: 10.9753/icce.v33.management.30. 7p. ISSN: 2156-1028

[11] Veloso-Gomes F, Costa J, Rodrigues A, Taveira-Pinto F, Pais-Barbosa J, Neves L. Costa da Caparica artificial sand nourishment and coastal dynamics. ICS2009 Proceedings. Journal of Coastal Research. 2009;**SI56**:678-682

[12] Antunes do Carmo JS. Climate change, adaptation measures and integrated coastal zone management: The new protection paradigm for the Portuguese coastal zone. Journal of Coastal Research. 2018;**34**(3):687-703. DOI: 10.2112/JCOASTRES-D-16-00165.1

[13] Guimarães MH, Dentinho T, Boski T. Application of the Q methodology in the promotion of dialogue among agents interested in the management of Praia da Vitória Bay, Terceira, Azores. In: VI Congress on Planning and Management of Coastal Areas of Portuguese Speaking Countries, Boavista Island, Cape Verde; 2011 (in Portuguese)

[14] Neumann B, Vafeidis AT, Zimmermann J, Nicholls RJ. Future coastal population growth and exposure to sea-level rise and coastal flooding—A global assessment. PLoS One. 2015;**10**(3):e0118571. DOI: 10.1371/journal.pone.0118571

[15] Berger M. Coastal Populations Grow—And Will Continue To—As Sea Levels Rise. Environment. 2015. Available from: https://weather.com/science/environment/news/coastal-populations-grow-sea-levels-rise

[16] EEA. The changing faces of Europe's coastal areas. EEA Report No. 6; Copenhagen; 2006. 112p

[17] Oh YI, Shin EC. Using submerged geotextile tubes in the protection of the E. Korean shore. Coastal Engineering. 2006;**53**:879-895. DOI: 10.1016/j.coastaleng.2006.06.005

[18] Taal MD, Löffler MAM, Vertegaal CTM, Wijsman JWM, Van der Valk L, Tonnon PK. Development of Sand Motor. Concise Report Describing the First Four Years of the Monitoring and Evaluation Programme (MEP). Delft, The Netherlands: Deltares; 2016. 62p

[19] Reis CS, Freitas H. Rehabilitation of the Leirosa sand dunes. In: EuroCoast-Portugal Association, editor. Littoral 2002, Porto; Vol. III. September 22-26, 2002. Porto, Portugal; 2002. pp. 381-384

[20] Reis CS, Freitas H, Antunes do Carmo JS. Leirosa sand dunes: A case study on coastal protection. In: Proceedings of the IMAM - Maritime Transportation and Exploitation of Ocean and Coastal Resources, Lisboa; de September 26-30. Taylor & Francis/BALKEMA; 2005. pp. 1469-1474. ISBN: Vol. 2: 0 415 39374 4; CD-Rom: 0 415 39433 3

[21] Antunes do Carmo JS, Reis CS, Freitas H. Successful rehabilitation of a sanddune system. WIT Transactions on Ecology and the Environment. 2006;**88**:195-204. DOI: 10.2495/CENV060191

[22] Antunes do Carmo JS, Reis CS, Freitas H. Rehabilitation of a geotextile-reinforced sand dune. Journal of Coastal Research, SI. 2009;**56**:282-286

[23] Antunes do Carmo JS, Reis CS, Freitas H. Working with nature by protecting sand dunes: Lessons learned. Journal of Coastal Research. 2010;**6**(6):1068-1078. DOI: 10.2112/JCOASTRES-D-10-00022.1

[24] Evans AJ, Garrod B, Firth LB, Hawkins SJ, Morris-Webb ES, Goudge H, et al. Stakeholder priorities for multi-functional coastal defence developments and steps to effective implementation. Marine Policy. 2017;**75**:143-155

[25] Mendonça A, Fortes CJ, Capitão R, Neves MG, Antunes do Carmo JS, Moura T. Hydrodynamics around an artificial surfing reef at Leirosa, Portugal. Journal of Waterway, Port, Coastal, and Ocean Engineering. 2012a;**138**(3):226-235. DOI: 10.1061/(ASCE)WW.1943-5460.0000128

[26] Mendonça A, Fortes CJ, Capitão R, Neves MG, Moura T, Antunes do Carmo JS. Wave hydrodynamics around a multi-functional artificial reef at Leirosa. Journal of Coastal Conservation. 2012;**16**:543-553. DOI: 10.1007/s11852-012-0196-1

[27] Lynett PJ, Liu PL-F. Modeling wave generation, evolution, and interaction with depth-integrated, dispersive wave equations. In: COULWAVE Code Manual. NY, USA: Cornell University; 2008. Long and intermediate wave modeling package, v. 2.0. https://pdfs.semanticscholar.org/f07d/a390c4590b880607037be8d900538f42c992.pdf

[28] Alveirinho Dias, J.M., Ferreira, Ó.M.F.C., Pereira A.P.R.R., 1994.

Estudo Sintético de Diagnóstico da Geomorfologia e da Dinâmica Sedimentar dos Troços Costeiros entre Espinho e Nazaré (e_book). Portugal. Available from: http://w3.ualg.pt/~jdias/JAD/eb_EspinhoNazare.html (in Portuguese) Portugal. Available from: http://w3.ualg.pt/~jdias/JAD/eb_EspinhoNazare.html

[29] Park H, Tomiczek T, Cox DT, van de Lindt JW, Lomonaco P. Experimental modeling of horizontal and vertical wave forces on an elevated coastal structure. Coastal Engineering. 2017;**128**:58-74. DOI: 10.1016/j.coastaleng.2017.08.001

[30] Coulbourne B, Haupt M, Sundberg S, Low DK, Yeung J, Squerciati J. Recommended residential construction for coastal areas. In: Building on Strong and Safe Foundations. 2nd ed. 2009. US Federal Emergency Management, FEMA P-550, 242 p. ISBN-13: 978-1484818657, ISBN-10: 1484818652

[31] Kamphuis JW. Introduction to coastal engineering and management. In: Advances Series on Ocean Engineering. 2nd ed. Vol. 30. Singapore: World Scientific Publishing; 2010. 525p. ISBN-13: 978-981-283-484-3

[32] IPCC. Synthesis report summary for policymakers. In: Core Writing Team, Pachauri RK, Meyer LA, editors. Climate Change 2014: Synthesis Report. Contribution of Working Groups I, II and III to the Fifth Assessment Report of the Intergovernmental Panel on Climate Change (IPCC); Geneva, Switzerland; 2014. 32p

[33] Jackson LA, Tomlinson R, Corbett B, Strauss D. Long term performance of a submerged coastal control structure: A case study of the Narrowneck multi-functional artificial reef. In: 33rd Conference on Coastal Engineering; Santander, Spain; Vol. 54; 2012. 13p

Chapter 2

Numerical Modeling Tools Applied to Estuarine and Coastal Hydrodynamics: A User Perspective

Isabel Iglesias, Paulo Avilez-Valente, José Luís Pinho, Ana Bio, José Manuel Vieira, Luísa Bastos and Fernando Veloso-Gomes

Abstract

Estuarine and coastal areas have been intensively studied given their complexity, ecological, and societal value and the importance of their ecosystem services. Estuarine and coastal management must be based on a sound characterization of these areas, which is achievable complementing the comprehensive field measurements with numerical models solutions. Based on a detailed comparison between two close-by, but extremely different, Portuguese estuaries (the Douro and Minho estuaries), this chapter intends to discuss how accurately numerical modeling tools can provide relevant information for a variety of coastal zones. They can be very useful for various applications in the planning and management fields, such as coastal and infrastructures protection, harbor activities, fisheries, tourism, and coastal population safety, thus supporting an effective and integrated estuarine and coastal management, which must consider both the safety of the populations and the sustainability of the marine ecosystems and services. In particular, the capacity of the numerical models to give a detailed characterization of morpho-hydrodynamic processes, as well as assess and predict the effects of anthropogenic interventions, extreme events and climate change effects, are presented.

Keywords: estuaries and coasts, hydrodynamics, field measurements, numerical modeling, coastal zones management

1. Introduction

The Coastal Zone is strategically important from environmental, economic, and societal points-of-view. Coastal zones are densely populated, concentrating human settlements, leisure activities, fisheries, and other marine industries. In the last decades, the population, economic assets, and urbanizations in the coastal zones have experienced a rapid growth, and a continuous increase of population in these regions is expected for the near future [1, 2]. The intensification of anthropic activities in coastal regions can boost their vulnerability to extreme events and, consequently, augment damages, cause injuries, and even loss of lives. In the present context of climate changes, an increase in the frequency and strength of extreme events have been reported [2], with potentially severe consequences for both society and environment, affecting human health and infrastructures, and resulting in the loss of property and habitats [3, 4]. For example, a loss of 70% of the coastal wetlands by 2080 is predicted due to a combination of sea-level rise, intensification of coastal urbanization, and increase of constructions for flood defense [5, 6]. Wetlands are highly productive areas, essential not only for fisheries and nature conservation, but also as a natural protection against floods. Their loss can have high socio-economic costs. Thus, extreme events, climate change conditions, and anthropogenic activities can put at risk the Coastal Zone's prolific flora and fauna and the ecosystem services they provide (e.g., food, fisheries, tourism, cultural services, energy, water abstraction, raw materials, water desalination/treatment, climate, and natural hazard regulation).

In this context, there is a need for scientific and technical information available to decision-makers, to support a sustainable coastal management, and avoid serious damages and higher losses for the littoral populations and coastal environments [7, 8]. This information is crucial to implement early warning systems and find solutions to reduce the negative impacts associated with extreme events (floods, droughts, storm surges, and coastal storms), climate change, and man-made interventions in the Coastal Zone. These can help to reduce exposure and vulnerability, mitigate the associated risks and promote the adaptation and the resilience of the communities to the potential adverse impacts, even though risks cannot be fully mitigated.

Effective protection of the Coastal Zone requires a comprehensive understanding of its morpho-hydrodynamic processes, as well as of the effects of these processes on the territory and the ecosystems. Coasts are land-ocean transition areas and, therefore, Coastal Zone assessment must consider land-ocean interactions, including estuaries and marine areas [9]. Meteorological, oceanographic, morphological, chemical, and biological parameters, obtained through *in-situ* measurement campaigns, are key descriptors to represent and understand the present state and the main evolution trends of the estuarine and coastal systems. However, field campaigns are usually expensive, often difficult and not always effective. As a consequence, there is a lack of continuous and long-term observations, being these regions generally under-sampled and poorly understood [10]. This limited knowledge, related with the lack of systematic monitoring and interface system complexity, leads to a high degree of uncertainty about the expectable effects of future scenarios associated with man-made interventions, climate change, and extreme events.

A complete estuarine/coastal dynamics characterization and, particularly, the assessment of future conditions, could be achieved through results obtained with numerical models [7, 11]. These models provide predictions of future trends and outcomes for different scenarios, hence supporting the implementation of sustainable action plans. Nevertheless, it should be noticed that field data is crucial for proper implementation of numerical models. Measured data are needed to define the models' initial states, forcing conditions, and static computation of calibration parameters values, or its dynamic computation using data assimilation techniques. Field data are also required to assess the numerical models performance comparing model results to measurements. So, despite the fact that advanced numerical models are excellent tools to understand the Coastal Zone behavior, comprehensive periodic or continuous monitoring campaigns in estuarine and coastal regions are still crucial to ensure the effectiveness of model application. Moreover, some of the complex morpho-hydrodynamic processes that take place in the coastal environment are still poorly understood. The combination of numerical and field monitoring methodologies should be pursued to improve our knowledge about those processes.

2. Numerical models and the *ensembles* technique

Considerable effort has been made to provide the most accurate estimations for the complex estuarine/coastal circulation, using either simple box models or complex numerical model suites. With the development of high-resolution numerical modeling systems, essential decision-making support instruments became available for an effective and integrated marine and coastal management. Numerical models are essential to properly assess the effect of each forcing driver, accurately representing the dynamical processes of estuarine/coastal systems [10]. Their input can be manipulated to represent the impact of changes in initial and boundary conditions, topo-bathymetric features, and coastal structures [12]. They can help to overcome the lack of field observations and measurements, allowing a full characterization of the morpho-hydrodynamic, chemical, and biological behavior of coastal regions, and providing valuable information to promote population, services and ecosystems safety [13, 14].

The current modeling tools available for coastal and estuarine studies allow an almost complete representation of the physical conditions of these areas. There is a large variety of models and techniques. The numerical techniques can be based on several methods, such as finite element, finite difference, finite volume, boundary element, or Eulerian-Lagrangian. The time integration algorithms can be explicit, implicit, semi-implicit, or characteristic-based. The functions can be of the first, second, or higher order, and the spatial dimensions can be one-dimensional (either in the horizontal plane 1DH or in the vertical 1DV), two-dimensional depth integrated (2DH) or lateral integrated (2DV), or three-dimensional (3D) [15, 16]. It is therefore important to properly select the adequate numerical model tool for the specific problem(s) the user wants to solve. This selection should be made in each case considering a compromise between the available data for model calibration and validation, the objectives of the model simulations and the available computational resources.

The most powerful modeling suites currently available (Delft3D, open TELEMAC-MASCARET, SWASH, ROMS, MOHID, SELFE, ADCIRC, Tuflow-FV, FVCOM, Mike21, etc.) are able to simulate several physical processes and environmental actions, such as flood/ebb cycles, bathymetry dynamics, friction, river discharge, water levels dynamics, currents velocity, wind action, waves, density effects, sediment transport, or Coriolis force, among others. Normally, numerical tools to model these processes are available in different and separate modules that the user can select depending on the desired complexity of the solutions. Modeling suites can also contain additional modules that allow the characterization of biological and ecological processes that are fundamental for water quality assessment, recurring to both Lagrangian and Eulerian transport approaches, including larvae migration, ecological status, nutrients concentration, pollutant evolution, etc. But, even these biogeochemical modules are completely dependent on the results obtained by the modules that represent the hydrodynamic patterns. The hydrodynamic conditions (water levels, currents velocities, temperature, and salinity) resulting, for example, from the complex interaction between tides, waves, storm surge, wave set-up, and river discharge, will define the main transportation patterns of sediments, larvae and pollutants with a direct effect on the ecosystems.

Most of the numerical models applied to coastal and estuarine regions can be implemented in 2DH/V or 3D configurations. 2DH simulations simplify the computational requirements by solving the shallow water equations. It has been demonstrated that this kind of models can accurately reproduce current velocity, flood extent, and water levels, being useful to complement risk assessment tools and early warning systems, because less computational resources are required and the numerical solutions are much faster obtained. Several authors used 2DH models with satisfactory results [14, 17–24], though 3D models are required to properly represent several processes, like vertical stratification, vertical current profiles, turbulent mixing processes, sediment transport, turbidity, water quality, effects of salinity, and temperature gradients on river plumes or salt-wedge estuarine configurations [25–28].

Regarding the equations and their approximations, 3D models are normally based on the Navier-Stokes equations or its depth-integrated version, the shallow water equations, which are applicable when the horizontal scale is much greater than the vertical scale. The shallow water equations applied to 1DH or 2DH problems are also known as Saint-Venant equations [29]. Despite their wide applicability, shallow water equations are not able to properly represent small relative amplitude waves propagating in shallow water conditions, which is of upmost importance to simulate the superposition between waves and currents, the effects of the waves on the sediment transport, the waves interaction with the bottom, or other wave processes, as shoaling, reflection, refraction, diffraction and decomposition [16, 30]. For this purpose, more sophisticated models are needed, like the ones of Boussinesq, Korteweg de Vries, Serre, or Green-Naghdi. These models include additional terms that take into account the nonhydrostatic effects of free surface curvature. Boussinesq equations [31, 32] are derived from the Navier-Stokes equations by depth-averaging them considering the pressure as nonhydrostatic [33]. Boussinesq-type equations can account nonlinear and dispersive effects considering various degrees of accuracy. The Korteweg de Vries equation, which was first introduced by Boussinesq and rediscovered by Korteweg and de Vries [32, 34] describes weakly nonlinear shallow water waves, allowing the representation of solitary wave solutions [35]. However, it must be considered that the wave dynamic becomes strongly nonlinear in the final stages of shoaling, particularly in the surf and swash zones. To properly represent this phenomenon and provide a correct description of the waves up to the breaking point, fully nonlinear equations should be considered, as the Serre or Green-Naghdi equations. Serre equations [36], also known as the Su-Gardner equations, are deduced from the fundamental equations of fluid mechanics, but taking into account the incompressibility of the fluid, the vertical uniformity of the velocity field and the conservation laws [16]. Finally, Green-Naghdi nonlinear equations [37] considers the 3D water-wave problem with a free surface and a variable bottom, and taking into account that the fluid can be rotational [38].

A comparison between the effectivity of different models and approximations is a really difficult task if the models are not implemented for the same region and considering the same initial/forcing conditions. Nevertheless, there are already some works where the capabilities of different numerical models are compared. Walstra et al. [39] applied the PISCES and the Delft3D models to coastal environments, depicting a general good overall agreement of both models, except for Delft3D under low wave conditions and for PISCES when a flow generated by the breaking of the waves on the shoal is presented. Rahman and Venugopal [25] compared 3D versions of open TELEMAC-MASCARET and Delft3D to represent the hydrodynamic conditions of the Pentland Firth and also the tidal regime energy in that area. Open TELEMAC-MASCARET seems to show the best consistency with the field data, although Delft3D also obtained good results for water level variations. The same models but in a 2DH version were selected by Iglesias et al. [14] to model floods at the Douro estuary. A small underestimation and overestimation of Delft3D and open TELEMAC-MASCARET, respectively, were observed for surface elevation for nonflood scenarios. For historical floods, the two models obtained very similar results, despite using different numerical approximations. The different numerical approximations of Mike 21 FM, Delft3D, and Delft3D FM were tested by Symonds et al. [24], confirming that despite the differences in the grids configuration, all the approaches accurately predict hydrodynamic conditions in complex estuarine regions. They also demonstrate that the unstructured models present a higher computational efficiency.

There is, hence, a wide range of numerical models that can be applied to estuarine/coastal zones to proper characterize these complex areas and gain a deep understanding of their hydrodynamic characteristics. Through the implementation of numerical algorithms, the circulation in these systems can be reproduced and different hydrodynamic processes can be represented. And knowledge about the hydrodynamic patterns influenced by bottom morphological changes may allow assessment and forecasting of the effects of hazardous and extreme events, anthropogenic intervention, or climate change. Hydrodynamic modeling has therefore been the focus of a large number of previous works in estuarine environments [11, 14, 17, 21, 22, 40–50].

Nowadays, the available powerful computational resources and complex numerical model suites allow the implementation of high-resolution studies with accurate results. Some examples are the works of Dias et al. [50] and Jones and Davies [46] for estuarine tides, Pinho and Vieira [44] for estuarine salt-water intrusion, Robins and Davies [21] and Sutherland et al. [43] for estuarine and coastal morpho-hydrodynamic behavior, Pinho et al. [45] and Iglesias et al. [14] for estuarine flood studies, Pinho et al. [51] for coastal waters hydrodynamics and water quality, Antunes do Carmo et al. [16] for agitation in harbors, Antunes do Carmo and Seabra-Santos [30] for coastal protection, and Monteiro et al. [22] for coastal circulation and river plumes. However, every modeling system has its own advantages and limitations, and model solutions will display uncertainties related with errors, calibration parameters, or model assumptions and forcing functions. Given the need for accurate forecasts, finding and implementing new solutions to avoid such errors is crucial. A single model can have biases, high variability, or inaccuracies related with the specification of initial conditions or the representation of physical processes in the models, causing large uncertainties in numerical prediction systems that can affect the reliability of the obtained results [52]. So, why not use several models to reduce uncertainties?

The *ensembles* technique is based on the combination of several numerical models solutions and can improve the forecast results when compared to a single model-based approach, even if the number of members in the models *ensemble* is as few as two [53]. *Ensemble* is a French word that means "set," "cluster," or "together" and usually refers to a unit or group of complementary parts that contribute to a single effect [54]. *Ensemble* modeling is a method that consists of running two or more related numerical models with the same conditions and then synthesizing the results into a single solution. This single solution will improve the accuracy of the final forecast, compared to the solutions provided by each model. It has been demonstrated that combining models generally increases the skill, reliability, and consistency of model forecasts [55], and *ensemble* simulations are currently applied in atmospheric and climate sciences to predict meteorological behavior and climate change effects ([56, 57]; IPCC: https://www.ipcc.ch/), being also used in other sectors, such as public health [58], agriculture [59], and more recently, in estuarine hydrodynamics [14].

Two types of *ensembles* can be defined: single-model and multi-model *ensembles*. The single-model *ensemble* is based on a single model that is run several times with very slight differences in the initial and/or boundary conditions, producing different simulation results. The multi-model *ensemble*, as its name suggests, considers different numerical models that present different structural complexities. Each model is run using the same initial and boundary conditions. The multi-model *ensemble* clearly outperforms both single models and the single-model *ensemble* [60–62].

To combine the outputs of different models, several statistical techniques can be applied, considering the mean, the median, a linear regression, weighted average, or linear programming techniques, among others [52, 57, 63–65]. Computing the mean of the *ensemble* members' outputs is one of the most used methods; using the mean, the effect of combining models to reduce errors can be expressed in terms of bias and variance [54]. The bias is used to measure the extent to which the *ensemble* result, averaged over the *ensemble* members, differs from the target function. The variance can be defined as the extent to which the considered *ensemble* models disagree [66]. Being the root mean square error (RMSE), the sum of the model variance, and square of the bias, the model's *ensemble* suitability can be expressed in terms of the mean RMSE of the *ensemble* members [14].

The *ensembles* technique can thus be used to improve single models forecasts, reduce their uncertainty, and provide the most accurate results for a variety of sectors, as previously mentioned. This is the main objective of the *EsCo-Ensembles* project, a FCT (*Fundação para a Ciência e a Tecnologia*, Portugal) funded project that intends to use a numerical models *ensemble* to simulate estuarine hydrodynamic patterns in the face of anthropic interventions, extreme events, and climate changes. In this project, the *ensembles* technique will be applied to two of the main estuaries of the Portuguese coast: the Douro and Minho estuaries.

3. A Portuguese case study: Douro and Minho estuaries

The need for an accurate forecasting based on numerical models becomes clear when the Douro and Minho estuarine regions are analyzed. Although these estuaries, located in the western coast of the Iberian Peninsula, are separated by a distance of less than 100 km (**Figure 1**), their dynamics and environmental conditions are completely different (**Table 1**). Despite presenting similar seasonal river flow regimes (**Figure 2**), with minimum values in summer and maximum values in the rainy winter season, flow values and patterns differ. For the Minho river, between 1970 and 2018, a maximum daily mean river flow of 4600 m^3/s was measured at Frieira dam; whereas the Douro presented a maximum daily mean river flow of 10,990 m^3/s at Crestuma-Lever dam between 1986 and 2018. The difference between these two systems is also evident when river flood peak discharges associated with different return periods are estimated (**Table 2**). The different river discharge patterns suffice to support the need for a focused and local hydrodynamic characterization of the two estuaries when trying to avoid future risks related with human interventions (dredging, ports, alterations in the estuarine banks, designing of breakwaters, etc.), extreme events such as coastal and estuarine floods or storm surges, sea-level rise and the increase in the number and/or intensity of extreme events associated with climate change predictions. However, a comparison of the main estuaries characteristics and bathymetric conditions (**Table 1** and **Figures 1** and **3**), reinforces this necessity. For the river flood peak discharges analysis, we assumed that the probability distribution for extreme flow events follows a Gumbel law [67]. Daily mean river flow data was provided by the SNIRH-*Sistema Nacional de Informação de Recursos Hídricos* (https://snirh.apambiente.pt/), measured between 1986 and 2018 at the Crestuma-Lever dam, and by *Confederación Hidrográfica Miño-Sil* (https://www.chminosil.es/es/), measured between 1970 and 2018 at the

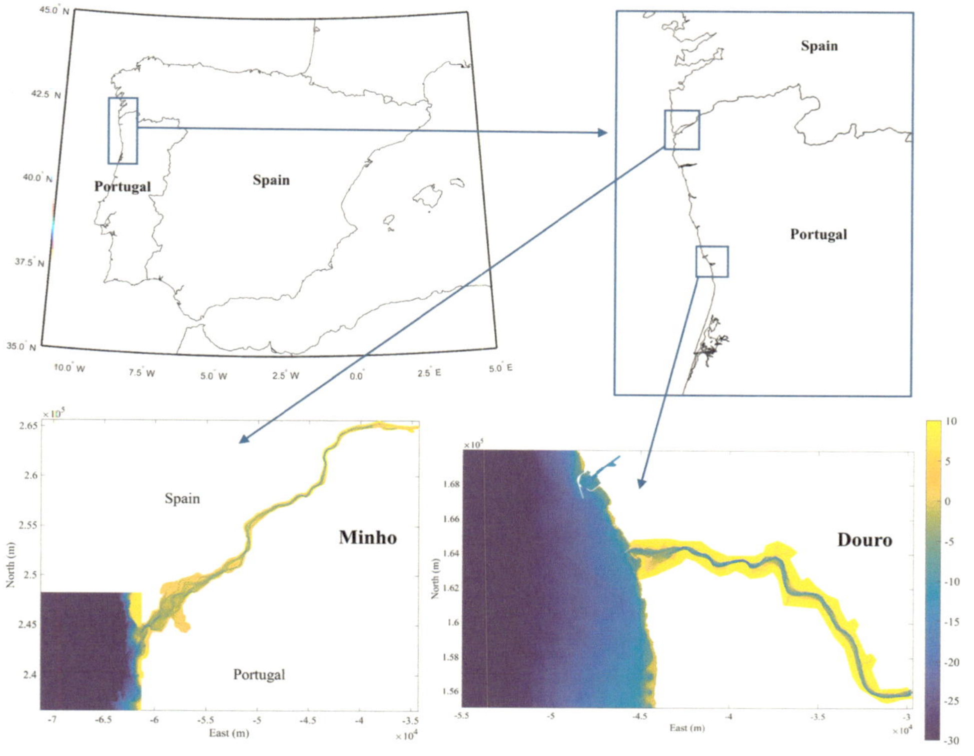

Figure 1.
Minho and Douro estuaries' location and bathymetry (in metres—vertical datum MSL-mean sea level). The horizontal coordinates of the bathymetric maps are in datum PT-TM06/ETRS89.

	Douro	Minho
Hydrographic basin (km^2)	97,603	17,080
River length (km)	927	300
Estuarine mouth	Artificial (breakwaters)	Natural (rocks)
Estuarine limit	Artificial (dam)	Natural (limit of tide penetration)
Estuarine extension (km)	21	35
Estuarine river flow	Artificial (dam)	Artificial (dam)
Mean depth (m)	−13.8	−7.6
Maximum estuarine width (m)	1,300	2,100
Minimum estuarine width (m)	135	160

Table 1.
Main characteristics of each considered estuary.

Frieira dam. It should be noted that the presented extreme values refer to average daily extremes, since annual instantaneous peak values were not available. Thus, higher flood peak extremes are expected for both estuaries.

The bathymetric map for the Minho estuary (**Figure 1**) was constructed using several topographic and bathymetric data sets: topographic data provided by the Portuguese Direção Geral do Território (DGT; http://www.dgterritorio.pt/) as a Digital Terrain Model (DTM) obtained from a nation-wide altimetric survey carried

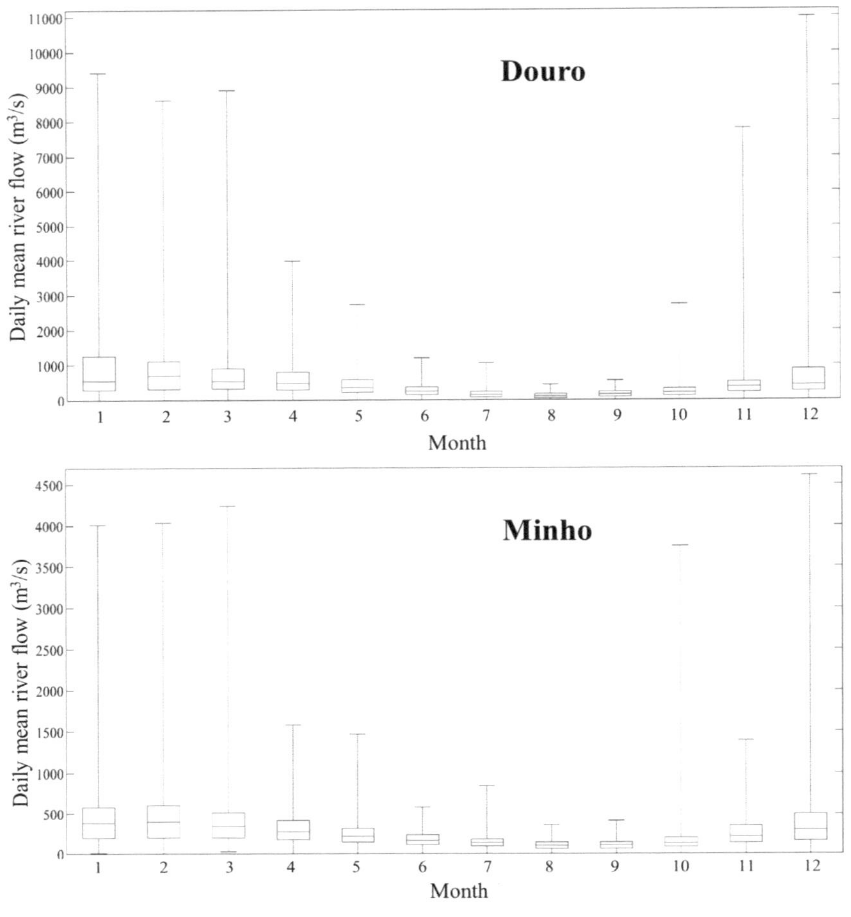

Figure 2.
Box plots representing the monthly minimum (lower whisker), lower quartile (25%, bottom box limit), median (red line), upper quartile (75%, top box limit), and maximum (upper whisker) daily mean river flow values of the Douro and Minho rivers flows, from 1986 to 2018 and from 1970 to 2018, respectively. Data source: SNIRH (https://snirh.apambiente.pt/) and Confederación Hidrográfica Miño-Sil (https://www.chminosil.es/es/).

T (years)	Q Douro (m^3/s)	Q Minho (m^3/s)
10	7,655	3,308
20	9,215	3,944
50	11,235	4,767
100	12,748	5,383
200	14,256	5,998
500	16,246	6,809
1000	17,749	7,421

Table 2.
River flood average daily extreme discharges (Q) associated with different return periods (T).

on in 2011 with a LiDAR (light detection and ranging); and bathymetric data provided by the Portuguese *Instituto Hidrográfico* (IH; http://www.hidrografico.pt/), obtained in 2006. The Douro bathymetry (**Figure 1**) was extracted from a 2009 bathymetric survey performed by the IH. The sand spit topography was taken from a 2015 topographical survey [68], and the adjacent coastal bathymetry was obtained from the Bathymetric Model of Douro (IH). For both estuaries, additional ocean bathymetric data was extracted from the GEBCO database [69], and the different data sets considered were interpolated using a Kriging algorithm [70, 71].

The Douro River is one of the major rivers of the Iberian Peninsula. It flows from the Sierra de Urbion, in Spain, to the Atlantic Ocean, in northern Portugal, ending in an urban estuary surrounded by two major cities: Porto and Vila Nova de Gaia (**Figure 1**). The Douro is a highly dynamic narrow estuary with torrential regimes that produce strong currents and recurrent severe floods that cause serious damage to the riverine populations and navigation problems [68, 72]. Its dynamics is mainly forced by freshwater flows, being very dependent on highly variable natural conditions and on the hydropower production schedule of the Crestuma-Lever dam and of the other 50 national and international river basin dams. For flow rates above 800 m^3/s, the river water masses rush to the sea and seawater intrusion is prevented. For flow rates below 800 m^3/s, the ocean water enters the estuary, which acquires a salt-wedge configuration [73]. This has a strong effect on the freshwater residence time, which can vary from 8 h to more than 2 weeks [74]. The bathymetric configuration of the Douro estuary presents an irregular distribution with depths generally varying between 0 and 10 m (**Figures 1** and **3**). Depths up to 28 m can be found associated with narrower sections, outer bends, and former sites of sediments extraction [72]. At the southern margin of the estuary's mouth lies a wetland (São Paio Bay) and an estuarine sand spit (Cabedelo) that partially obstructs the river mouth, protecting the estuary from the ocean's storm waves. This sand spit is made up of fluvial and maritime sediments, and its morphodynamics is conditioned by natural (wind, rainfall, river flow, waves, tides and storm surges) and human (breakwaters and dams construction, sand extraction, and dredging) processes [68, 75, 76]. To prevent erosion of the sand spit or its excessive migration into the navigation channel, maintaining both navigability and bank protection, a new detached breakwater was built parallel to the head of the sandbar, and the existing northern breakwater was extended. These structures, concluded in 2008, interfere with local sedimentary and hydrodynamic patterns, significantly increasing the area and volume of the sand spit in a relatively short period ($\sim$10 years) [68, 77]. Historical records reveal ruptures or partial destruction of the sandbar during river flood events, allowing for a rapid discharge of excess

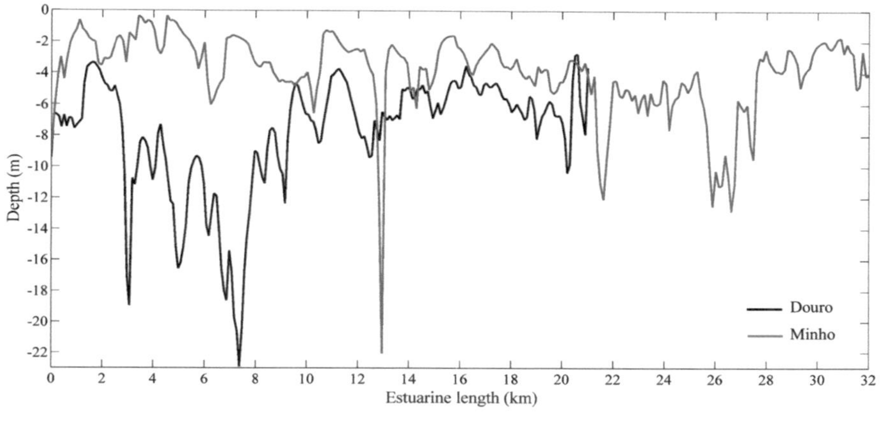

Figure 3.
Minho and Douro estuaries' bathymetric profile for the estuarine central axis (in metres—vertical datum MSL-mean sea level).

water and reducing the risk of urban flooding [68]. Now, with a stronger sandbar, its rupture is less probable and the effect of a flood is likely to be harsher, both in terms of economic losses and structural damage [14].

There are some previous works that aimed to reproduce the Douro estuary dynamics using numerical models. Silva [78] implemented a 2D depth-averaged (2DH) hydrodynamic model for the lower estuarine area, to represent the effect of several engineering works on the main currents configuration. Using the 1994 configuration, he found that relatively small alterations of the estuarine mouth conditions can produce marked changes in the currents strength and direction for normal winter conditions (river flow $\sim$1000 m^3/s). The impact of the structures at the estuary's mouth on hydrodynamics, salt-water intrusion, and sediment transport was also presented by Pinho et al. [79], considering several scenarios of mean river flow. Their solutions, which were obtained using two coupled models: one for hydrodynamics and another for sediment transport, revealed maximum current velocities and maximum erosion at the estuarine mouth between the breakwaters. Similar patterns were obtained by Portela [72] and Iglesias et al. [14]. Particularly, Iglesias et al. [14] implemented two different numerical models for the Douro river to depict the effect of the sand spit in the floods water levels, demonstrating that the new breakwaters configuration and the strengthening of the sand spit will probably produce an increase in high-water levels during flood conditions, with expected severe impacts on the estuary banks. River floods were also simulated by Araújo et al. [80]. However, their work was focused on numerical model meshes development rather than on the socio-economical impacts of floods. Other related research are the modeling works of Azevedo et al. [81, 82], which related the estuary's hydrodynamic behavior with contaminant dispersion, biogeochemistry, and primary production, and the work of Mendes et al. [83], which evaluated the potential effect of sea-level rise in the Douro estuary.

The Minho River is an international river as well; it separates Spain and Portugal in its last 70 km, flowing into the Atlantic Ocean between A Guarda (Spain) and Caminha (Portugal). The Minho estuary is a very shallow water body, with a mean depth of 4 m, but regions close to 20-m depth can be found associated with a narrowing of the main channel, that increases flow velocities and, consequently, erosion (**Figures 1** and **3**) [84, 85]. Estuarine flows are mainly controlled by the Frieira dam, whose reservoir feeds a hydroelectric power plant located 80 km upstream from the estuary mouth. Due to the low river flow values (see **Figure 2**), the average water residence time in this estuary is 1.5 days [86]. The tributaries of the Minho between Frieira and the estuarine mouth can provide some additional freshwater to the estuary, but they have a minor influence given their small drainage basins.

One of the most important characteristics of the Minho estuarine region is its large diversity of habitats and its importance for the nursery and feeding of marine species and for ecosystem functioning [87, 88]. For this reason, this estuary is protected by Portuguese and Spanish conservation statutes, preserving a low level of industrialization. Despite the fact that the Minho ecosystem has been intensively studied in terms of its morpho-hydrodynamic characteristics, water quality, biodiversity, populations, and pollution, its dynamics is still essentially unknown [89, 90]. One of this estuary's main problems is the strong siltation related with high sediment deposition and low currents velocities, which are due to flow rate smoothing by the dam, and the consequent reduction of the frequency and intensity of floods [91, 92]. Being the hydrographic zero the level of the lowest astronomical tide, which is 2 m below the local mean sea level, the area above the hydrographic zero between the river mouth, and 14 km upstream represents about 70% of the total area, indicating a high degree of sedimentation. The morphodynamic patterns generated by silting produced several bathymetric constraints to navigation, such as strangulation or intense variations of bathymetry, and various islands and sandbars that emerge during low tide [85, 93]. Also, during the low water level period of spring tides, the connection between the estuary and the sea is restricted to two shallow channels, causing serious problems for navigation. Dredging campaigns are often carried out to keep the navigation channel open, with possible implications for the morphological evolution of the estuary and adjacent coastal areas [94], and for bottom habitats.

In terms of numerical modeling, there is a lack of publications for this estuary, probably due to the scarcity of *in-situ* data to force and calibrate the numerical models and validate their accuracy. Sousa et al. [95] and Pereira [13] implemented two hydrodynamic numerical models (MOHID and Delft3D) for the Minho estuary surroundings. In spite of obtaining interesting results in the validation processes with data from the Minho estuary measurement stations, those studies focused on representing the interaction between the waters of the Minho river and the Galician rias or the Lima estuary, respectively, and do not present a detailed and complete analysis of the estuarine hydrodynamics. Delgado [94] and Portela [92] also built some numerical modeling tools for the Minho estuary. Their studies focused mainly on the estuarine sediment transport for a few theoretical scenarios.

Nevertheless, several conclusions about the morpho-hydrodynamic behavior of the Minho estuary can be inferred from previous works. The estuarine processes may be dominated by the river flow or by the tide depending on the magnitude of these two forcing parameters. Extreme river flows can change the circulation pattern within the entire estuarine and coastal region, restricting the entrance of oceanic water to the mouth of the estuary. In such cases, the tide acts as a resistance to the fluvial flow. Current velocities of the ebb are higher than the velocity during flood, which produces a higher duration of the ebb. This effect is stronger for low river flows. As expected, currents exhibit higher velocities in the narrower sections, particularly at the mouth of the estuary, but they are also stronger during spring tides than during neap tides. The described hydrodynamic patterns will have a direct effect on sediment transport, which is directly proportional to the strength of the flow and the amplitude of the tide. Similar conclusions were obtained by Iglesias et al. [96] using realistic river flow and tide scenarios. Their numerical solutions show a tide dominated estuary, with a visible tidal effect even for extreme river flows. For low river flow conditions, a large part of the estuarine region is dry, becoming exposed to the wind action. In this situation, river flow is confined to two shallow channels in the estuarine area. During high flow conditions, most of the estuary is flooded, with intense currents throughout the estuarine region, except in the widest part upstream from the river mouth, where the estuary widens and the cross-sectional area increases significantly.

4. Discussion and conclusions

The characteristics of an estuarine region depend on numerous drivers that define not only the estuarine behavior, but also its ecosystems and human settlement distribution. In this chapter, these relationships, and differences between estuaries, are highlighted, comparing two near-by but completely different estuarine regions of the northern Portuguese coast. The comparison reinforces the need to analyze each region separately, considering the specific configuration of each estuary for the definition of management protocols, to minimize any potential vulnerability and to allow mitigation of risks and hazardous effects.

The Douro estuary is an urban estuary, where effects of extreme events, anthropic activities, and climate changes will mostly generate problems in urban environments, reflected in structural damages, economic losses, and impacts on tourism and navigation activities. In the Minho estuary, the same phenomena will have different impacts, mostly affecting ecosystems and biodiversity, due to habitat loss and the migration and loss of autochthonous species. In comparison to the Douro estuary, the Minho presents a smaller concentration of population and human activities on its banks. Thus, economic impacts will mainly be caused by changes in the fishing and tourism activities.

The modeling tools that have been developed so far, and which were described above, although extremely useful, need to be further developed. In the case of the Douro estuary, a complete hydrodynamic characterization, which considers the present topo-bathymetric configuration is still needed to fully unravel the estuarine dynamics, assess evolution trends, forecast future developments, including the effects of possible future human interventions, as well as to estimate the risks of flooding and the effect of sea-level rise associated with global warming. The numerical models developed for the Minho estuary are clearly insufficient for a complete characterization of this complex region, as well as for a reliable forecast of, for instance, hazardous events effects. Numerical models, capable of representing the estuarine stratification and its links with tide, river flow, and waves, are key tools to understand the distribution of biota and the functioning of the ecosystem, and to anticipate possible future conditions considering climate change and sea-level rise conditions. In addition, numerical models capable of representing the transport of larvae, pollutant, and sediments in a realistic way are crucial to describe estuarine trends, assess effects of anthropogenic intervention, quantify water residence times and sedimentary/erosive processes, as well as anticipate the effect of extreme river flows.

Next to stressing the relevance of performing regional modeling studies, this chapter also provides a thorough characterization of different available models, techniques, and physical processes simulations, including comparisons of several models performances, underlining the importance of choosing the modeling tool that best suits the numerical problem at hand and the computational means available to the user.

All these facts highlight the relevance of research projects dedicated to improve numerical modeling tools that provide a deeper understanding of the estuarine and coastal zones and represent the dynamics of the systems over time (past and present situations). One example is the project *EsCo-Ensembles—Estuarine and coastal numerical modeling ensembles for anthropogenic, extreme events and climate change scenarios*. This project aims to apply a new methodology based on models *ensembles* to build forecast and warning systems for estuarine/coastal regions. Two or more model outputs will be combined to obtain more accurate results that properly represent actual estuarine behavior as well as future trends, allowing for identification and mapping of the most sensitive areas. Modeling results will help to preserve and protect the estuarine/coastal regions and to mitigate the damages related with hazardous events, anthropogenic interventions, and climate change. Although extremely reliable, numerical models are prone to errors related to bathymetric and topographic assumptions, grid construction, spatial and temporal resolution, and to initial and forcing conditions, as well. As an example, there is not a clear trend for sea-level rise at the Portuguese coast. For different locations and periods, several authors report values of 1.3 [97], 1.9 [98], or even −0.7 mm/year [99], which does not agree with the global mean sea level rise of 1–2 mm/year, although locally, sea level can present increasing or decreasing trends [2]. To properly represent the effect of sea-level rise over the hydrodynamic conditions for a particular region, all these trends should be carefully evaluated and properly integrated into the numerical models. However, to avoid a large amount of input options, which can lead to inaccuracies in the numerical results, a models *ensemble* will be produced for the final forecasting. The *EsCo-Ensembles* project will provide valuable information about the NW Portuguese coastal zone to authorities, stakeholders, inhabitants, and society in general. Project results will contribute to the development of strategies for a sustainable management of estuarine and coastal areas affected by anthropogenic pressures, providing key information to develop protocols and mitigation strategies, to protect natural resources, population and infrastructures, and to potentiate new maritime infrastructures and coastal defense works adapted to future harmful events and climate change conditions.

Acknowledgements

This research was partially supported by the Strategic Funding UID/Multi/04423/2013 through national funds provided by FCT—Foundation for Science and Technology and European Regional Development Fund (ERDF). This contribution has also been funded by project EsCo-Ensembles (PTDC/ECI-EGC/30877/2017), co-funded by NORTE2020, Portugal 2020, and the European Union through the ERDF, and by FCT through national funds.

Author details

Isabel Iglesias[1*], Paulo Avilez-Valente[1,2], José Luís Pinho[3], Ana Bio[1], José Manuel Vieira[3], Luísa Bastos[1,4] and Fernando Veloso-Gomes[1,2]

1 Interdisciplinary Centre of Marine and Environmental Research (CIIMAR), University of Porto, Matosinhos, Portugal

2 Faculty of Engineering, University of Porto (FEUP), Porto, Portugal

3 Center of the Territory, Environment and Construction (CTAC), University of Minho, Braga, Portugal

4 Faculty of Sciences, University of Porto (FCUP), Porto, Portugal

*Address all correspondence to: iiglesias@ciimar.up.pt

References

[1] Dangendorf S, Wahl T, Hein H, Jensen J, Mai S, Mudersbach C. Mean sea level variability and influence of the North Atlantic oscillation on long-term trends in the German bight. Water. 2012;**4**(4):170-195

[2] IPCC. Managing the risks of extreme events and disasters to advance climate change adaptation. Summary for Policymakers. Field C, Barros V, Stocker T, Qin D, Dokken D, Ebi K, et al., editors. Cambridge, UK, and New York, NY, USA: Cambridge University Press; 2012. 3-21 p

[3] Vose RS, Applequist S, Bourassa MA, Pryor SC, Barthelmie RJ, Blanton B, et al. Monitoring and understanding changes in extremes: Extratropical storms, winds, and waves. Bulletin of the American Meteorological Society. 2014;**95**(3):377-386

[4] Bell JE, Brown CL, Conlon K, Herring S, Kunkel KE, Lawrimore J, et al. Changes in extreme events and the potential impacts on human health. Journal of the Air & Waste Management Association. 2017;**64**(4): 265-287

[5] Nicholls RJ, Hoozemans FMJ, Marchand M. Increasing flood risk and wetland losses due to global sea-level rise: Regional and global analyses. Global Environmental Change. 1999; **9**(1):S69-S87

[6] Nicholls RJ. Coastal flooding and wetland loss in the 21st century: Changes under the SRES climate and socio-economic scenarios. Global Environmental Change. 2004;**14**(1): 69-86

[7] Coelho C, Silva R, Veloso-Gomes F, Taveira-Pinto F. Potential effects of climate change on northwest Portuguese coastal zones. ICES Journal of Marine Science. 2009;**66**:1497-1507

[8] Veloso-Gomes F. Zonas costeiras: Variabilidade climática, ações antropogénicas e alterações climáticas. In: II Congresso Ibero-Americano de Gestão Integrada de Áreas Litorais. 2016. pp. 757-758

[9] Veloso-Gomes F, Barroco A, Ramos Pereira A, Sousa Reis C, Calado H, Gomes Ferreira J, et al. Basis for a national strategy for integrated coastal zone management—in Portugal. Journal of Coastal Conservation. 2008;**12**(1):3-9

[10] Bastos L, Bio A, Iglesias I. The importance of marine observatories and of RAIA in particular. Frontiers in Marine Science. 2016;**3**:1-11

[11] Iglesias I, Venancio S, Peixoto R, Pinho J, Avilez-Valente P, Vieira J. The Douro estuary: Modelling comparison for floods prevention. In: Actas das 4.as Jornadas de Engenharia Hidrográfica. Lisbon: Instituto Hidrográfico; 2016. pp. 159-162

[12] Teng J, Jakeman AJ, Vaze J, Croke BFW, Dutta D, Kim S. Flood inundation modelling: A review of methods, recent advances and uncertainty analysis. Environmental Modelling and Software. 2017;**90**:201-216

[13] Pereira H. Coupled modelling of the Minho and Lima estuaries: Hydrological response to climate changes [Thesis]. Universidade de Aveiro; 2016

[14] Iglesias I, Venâncio S, Pinho JL, Avilez-Valente P, Vieira JMP. Two models solutions for the Douro estuary: Flood risk assessment and breakwater effects. Estuaries and Coasts. 2018; **42**(2):348-364

[15] Panayotis P. Modelling coastal hydrodynamics [Internet]. 2016. Available from: http://www.coastalwiki.org/wiki/Modelling_coastal_hydrodynamics [Accessed: 2019-02-14]

[16] Antunes do Carmo JS, Seabra Santos FJ, Almeida AB. Numerical solution of the generalized Serre equations with the MacCormack finite difference scheme. International Journal for Numerical Methods in Fluids. 1993;**16**:725-738

[17] Horritt MS, Bates PD. Evaluation of 1D and 2D numerical models for predicting river flood inundation. Journal of Hydrology. 2002;**268**(1–4): 87-99

[18] Hu K, Ding P, Wang Z, Yang S. A 2D/3D hydrodynamic and sediment transport model for the Yangtze estuary, China. Journal of Marine Systems. 2009;**77**(1–2):114-136

[19] Néelz S, Pender G. Benchmarking of 2D Dyfraulic modelling packages [Internet]. 2010. Available from: https://www.gov.uk/government/uploads/system/uploads/%0Aattachment_data/file/290884/scho0510bsno-e-e.pdf [Accessed: 2019-02-14]

[20] Néelz S, Pender G. Benchmarking the latest generation of 2D hydraulic modelling packages [Internet]. 2013. Available from: http://evidence.environment-agency.gov.uk/%0AFCERM/Libraries/FCERM_Project_Documents/SC120002_%0ABenchmarking_2D_hydraulic_models_Report.sflb.ashx [Accessed: 2019-02-14]

[21] Robins PE, Davies AG. Morphological controls in sandy estuaries: The influence of tidal flats and bathymetry on sediment transport. Ocean Dynamics. 2010;**60**(3):503-517

[22] Monteiro IO, Marques WC, Fernandes EH, Gonçalves RC, Möller OO. On the effect of earth rotation, river discharge, tidal oscillations, and wind in the dynamics of the Patos lagoon coastal plume. Journal of Coastal Research. 2011;**27**(1988):120-130

[23] Wan Y, Gu F, Wu H, Roelvink D. Hydrodynamic evolutions at the Yangtze estuary from 1998 to 2009. Applied Ocean Research. 2014;**47**: 291-302

[24] Symonds AM, Vijverberg T, Post S, van der Spek B, Henrotte J, Sokolewicz M. Comparison between Mike 21 FM, Delft3D and Delft3D FM flow models of Western Port Bay, Australia. In: Proceedings of 35th Conference on Coastal Engineering, Antalya, Turkey. 2016

[25] Rahman A, Venugopal V. Inter-comparison of 3D tidal flow models applied to Orkney Islands and Pentland Firth. In: 11th Eur Wave Tidal Energy Conf (EWTEC 2015). 2015. pp. 1-10

[26] Cancino L, Neves R. Hydrodynamic and sediment suspension modelling in estuarine systems part II: Application to the Western Scheldt and Gironde estuaries. Journal of Marine Systems. 1999;**22**(2–3):117-131

[27] Kärnä T, Baptista AM, Lopez JE, Turner PJ, McNeil C, Sanford TB. Numerical modeling of circulation in high-energy estuaries: A Columbia River estuary benchmark. Ocean Modelling. 2015;**88**:54-71

[28] Marques WC, Fernandes EH, Monteiro IO, Möller OO. Numerical modeling of the Patos lagoon coastal plume, Brazil. Continental Shelf Research. 2009;**29**(3):556-571

[29] Miglio E, Perotto S, Saleri F. A multiphysics strategy for free surface flows. In: Chapter in Lecture Notes in Computational Science and Engineering. 2015. DOI: 10.1007/3-540-26825-1-40

[30] Antunes do Carmo JS, Seabra Santos FJ. Near-shore sediment dynamics computation under the combined effects of waves and currents. Advances in Engineering Software. 2002;**33**:37-48

[31] Boussinesq JV. Théorie des ondes et des remous qui se propagent le long d'un canal rectangulaire horizontal, en communiquant au liquide contenu dans ce canal des vitesses sensiblement pareilles de la surface au fond. Journal de Mathématiques Pures Appliquées Deuxième Série. 1872;**17**:55-108

[32] Boussinesq JV. Essai sur la théorie des eaux courantes. Mémoires présentés par Divers à l'Acad des Sci Inst Nat Fr. 1877;XXIII:1–680.

[33] Hervouet JM, Couche P, Guesmia M. Solving Boussinesq equations in finite elements. Transactions on the Built Environment. 1999;**40**:213-221

[34] Korteweg DJ, de Vries G. On the change of form of long waves advancing in arectangular canal, and on a new type of long stationary waves. Philosophical Magazine. 1895;**36**(5):422-443

[35] Weisstein EW. Korteweg-de Vries Equation [Internet]. Available from: http://mathworld.wolfram.com/Korteweg-deVriesEquation.html [Accessed: 019–02-14]

[36] Serre F. Contribution à l'étude des écoulements permanents et variables dans les canaux. La Houille Blanche. 1953;**8**:374-388

[37] Green AE, Naghdi PM. A derivation of equations for wave propagation in water of variable depth. Journal of Fluid Mechanics. 1976;**78**:237-246

[38] Ionescu-Kruse D. Variational derivation of the Green-Naghdi shallow-water equations. Journal of Nonlinear Mathematical Physics. 2012; **19**:1-19

[39] Walstra DJR, Sutherland J, Hall L, Blogg H, van Ormondt M. Verification and comparison of two hydrodynamic area models for an inlet system. Proceedings of the 4th Coastal Dynamics Conference Lund, Sweden. 2001;**40566**:433-442

[40] Iglesias I, Couvelard X, Avilez-Valente P, Caldeira R. NW Iberia shelf dynamics: The river Douro plume. In: Book of Proceedings 3rd IAHR Europe Congress. IAHR: Porto; 2014

[41] Mesquita A, Avilez-Valente P, Tavares-Pinho F. Turbulência e rebentação sobre quebramares submersos. In: Proceedings of Congreso de Métodos Numéricos en Ingeniería. 2009

[42] Peixoto J, Pinho J. Modelação hidrodinâmica do estuário do rio Douro. In: Proceedings of Congreso de Métodos Numéricos en Ingeniería. 2013. pp. 25-28

[43] Sutherland J, Walstra DJR, Chesher TJ, van Rijn LC, Southgate HN. Evaluation of coastal area modelling systems at an estuary mouth. Coastal Engineering. 2004;**51**(2):119-142

[44] Pinho J, Vieira J. Mathematical modelling of salt water intrusion in a northern Portuguese estuary. In: Ferreira L, Vieira J, editors. Water in Celtic Countries: Quantity, Quality and Climate Variability. IAHR; 2007. pp. 277-287

[45] Pinho J, Ferreira R, Vieira L, Schwanenberg D. Comparison between two hydrodynamic models for flooding simulations at river Lima basin. Water Resoururces Management. 2015;**29**(2): 431-444

[46] Jones J, Davies AM. Application of a finite element model to the computation of tides in the Mersey estuary and eastern Irish Sea. Continental Shelf Research. 2010;**30**(5):491-514

[47] van Maren DS, van Kessel T, Cronin K, Sittoni L. The impact of channel deepening and dredging on estuarine sediment concentration. Continental Shelf Research. 2015;**95**:1-14

[48] Gomes MP, Pinho JL, Antunes do Carmo JS, Santos L. Hazard assessment

of storm events for the battery, New York. Ocean and Coastal Management. 2015;**118**:22-31

[49] Avilez-Valente P, Seabra-Santos FJ. A high-order Petrov-Galerkin finite element method for the classical Boussinesq wave model. International Journal for Numerical Methods in Fluids. 2009;**59**(9):969-1010

[50] Dias JM, Valentim JM, Sousa MC. A numerical study of local variations in tidal regime of Tagus estuary, Portugal. PLoS One. 2013;**8**(12):e80450

[51] Pinho JLS, Vieira JMP, Antunes do Carmo JS. Hydroinformatic environment for coastal waters hydrodynamics and water quality modelling. Advances in Engineering Software. 2004;**35**:205-222

[52] Rozante J, Moreira D, Godoy R, Fernandes A. Multi-model ensemble: Technique and validation. Geoscientific Model Development. 2014;**7**:2333-2343

[53] Baker L, Ellison D. Optimisation of pedotransfer functions using an artificial neural network ensemble method. Geoderma. 2008;**144**(1–2): 212-224

[54] Mohan Das D, Singh R, Kumar A, Mailapalli DR, Mishra A Chatterjee CA. Multi model ensemble approach for stream flow simulation. In: Panigrahi B, Goyal MR, editors. Innovations in Agricultural and Biological Engineering, Modeling Methods and Practices in Soil and Water Engineering. Apple Academic Press; 2016; Chapter 4. pp. 71-102

[55] Tebaldi C, Knutti R. The use of the multi-model ensemble in probabilistic climate projections. Philosophical Transactions of the Royal Society A - Mathematical Physical and Engineering Sciences. 2007;**365**:2053-2075

[56] Weigel AP, Liniger MA, Appenzeller C. Can multi-model combination really enhance the prediction skill of probabilistic ensemble forecasts? Quarterly Journal of the Royal Meteorological Society. 2007; **134**:241-260

[57] WMO. Guidelines on Ensemble Prediction Systems and Forecasting [Internet]. World Meteorological Organization, Geneva; 2012. Available from: http://www.wmo.int/pages/prog/www/Documents/1091_en.pdf [Accessed: 2019-02-14]

[58] Thomson MC, Doblas-Reyes FJ, Mason SJ, Hagedorn R, Connor SJ, Phindela T, et al. Malaria early warnings based on seasonal climate forecasts from multi-model ensembles. Nature. 2006; **439**:576-579

[59] Cantelaube P, Terres JM. Seasonal weather forecasts for crop yield modelling in Europe. Tellus A: Dynamic Meteorology and Oceanography. 2005; **57**:476-487

[60] Georgakakos KP, Seo DJ, Gupta H, Schaake J, Butts MB. Towards the characterization of streamflow simulation uncertainty through multimodel ensembles. Journal of Hydrology. 2004;**298**:222-241

[61] Krishnamurti TN, Kishtawal CM, LaRow TE, Bachiochi DR, Zhang Z, Eric Williford C, et al. Improved weather and seasonal climate forecasts from multimodel superensemble. Science. 1999;**285**:1548-1550

[62] Palmer TN, Doblas-Reyes FJ, Hagedorn R, Weisheimer A. Probabilistic prediction of climate using multi-model ensembles: From basics to applications. Philosophical Transactions of the Royal Society, B: Biological Sciences. 2005;**360**:1991-1998

[63] Roy Bhowmik SK, Durai VR. Multi-model ensemble forecasting of rainfall over Indian monsoon region. Atmosfera. 2008;**21**:225-239

[64] Roy Bhowmik SK, Durai VR. Application of multimodel ensemble techniques for real time district level rainfall forecasts in short range time scale over Indian region. Meteorology and Atmospheric Physics. 2010;**106**: 19-35

[65] Ajami NK, Duan Q, Gao X, Sorooshian S. Multimodel combination techniques for analysis of hydrological simulations: Application to distributed model Intercomparison project results. Journal of Hydrometeorology. 2006;7: 755-768

[66] Krogh A, Vedelsby J. Neural network ensembles, cross validation, and active learning. In: Advances in Neural Information Processing Systems 7. Vol. 7. 1995. pp. 231-238

[67] Loaiciga HA, Leipnik RB. Analysis of extreme hydrologic events with Gumbel distributions: Marginal and additive cases. Sthocastic Environmental Research and Risk Assessment. 1999;**13**:251-259

[68] Bastos L, Bio A, Pinho JLS, Granja H, Jorge da Silva A. Dynamics of the Douro estuary sand spit before and after breakwater construction. Estuarine, Coastal and Shelf Science. 2012;**109**: 53-69

[69] Becker JJ, Sandwell DT, Smith WHF, Braud J, Binder B, Depner J, et al. Global bathymetry and elevation data at 30 arc seconds resolution: SRTM30_ PLUS. Marine Geodesy. 2009;**32**(4): 355-571

[70] Krige DG. A Statistical Approach to some Mine Valuations and Allied Problems at the Witwatersrand [Thesis]. University of Witwatersrand; 1951

[71] Matheron G. Principles of geostatistics. Economic Geology. 1963; **58**:1246-1266

[72] Portela LI. Sediment transport and morphodynamics of the Douro River estuary. Geo-Marine Letters. 2008; **28**(2):77-86

[73] Azevedo IC, Duarte PM, Bordalo AA. Understanding spatial and temporal dynamics of key environmental characteristics in a mesotidal Atlantic estuary (Douro, NW Portugal). Estuarine, Coastal and Shelf Science. 2008;**76**(3):620-633

[74] Vieira MEC, Bordalo AA. The Douro estuary (Portugal): A mesotidal salt wedge. Oceanologica Acta. 2000;**23**(5): 585-594

[75] Santos I, Teodoro A, Taveira-Pinto F. Análise da evolução morfológica da restinga do rio Douro. In: 5as Jornadas de Hidráulica, Recursos Hídricos e Ambiente. Vol. 2010. Porto. p. 14

[76] Granja H, Bastos L, Pinho J, Gonçalves J, Henriques R, Bio A, et al. Integração de metodologias no estabelecimento de um programa de monitorização costeira para avaliação de risco. In: VII Conferência Nacional de Cartografia e Geodesia. 2011. p. 11

[77] Teodoro AC, Pais-Barbosa J, Gonçalves H, Veloso-Gomes F, Taveira-Pinto F. Extraction of Cabedelo sand spit area (Douro estuary) from satellite images through image processing techniques. Journal of Coastal Research. 2011;**64**(64):1740-1744

[78] Silva A. Implementação de um modelo hidromorfológico para a Barra do Douro: Contribuição para a compreensão do sistema. In: 3° Congresso da Água. Lisboa; 1996

[79] Pinho J, Vieira J, Neves D. Efeito das obras da embocadura na hidrodinâmica, intrusão salina e dinâmica sedimentar do estuário do Rio Douro. In: 10° Congresso da Água. Alvor. 2010

[80] Araújo MAVC, Mazzolari A, Trigo-Teixeira A. An object oriented mesh generator: Application to flooding in the Douro estuary. Journal of Coastal Research. 2013;**65**(65):642-647

[81] Azevedo IC, Bordalo AA, Duarte PM. Influence of river discharge patterns on the hydrodynamics and potential contaminant dispersion in the Douro estuary (Portugal). Water Research. 2010;**44**(10):3133-3146

[82] Azevedo IC, Bordalo AA, Duarte P. Influence of freshwater inflow variability on the Douro estuary primary productivity: A modelling study. Ecological Modelling. 2014;**272**: 1-15

[83] Mendes R, Vaz N, Dias JM. Potential impacts of the mean sea level rise on the hydrodynamics of the Douro river estuary. Journal of Coastal Research. 2013;**165**(65):1951-1956

[84] Freitas V, Costa-Dias S, Campos J, Bio A, Santos P, Antunes C. Patterns in abundance and distribution of juvenile flounder, Platichthys flesus, in Minho estuary (NW Iberian Peninsula). Aquatic Ecology. 2009;**43**(4):1143-1153

[85] Reis JL, Martinho AS, Pires-Silva AA, Silva AJ. Assessing the influence of the river discharge on the Minho estuary tidal regime. Journal of Coastal Research. 2009;**2009**(56):1405-1409

[86] Ferreira J, Abreu P, Bettencourt A, Bricker S, Marques J, Melo J, et al. Monitoring Plan for Portuguese Coastal Waters. Water Quality and Ecology. Development of Guidelines for the Applications of the European Union Water Framework Directive; 2005. p. 141

[87] Domínguez García MD, Horlings L, Swagemakers P, Simón Fernández X. Place branding and endogenous rural development. Departure points for developing an inner brand of the river Minho estuary. Place Branding and Public Diplomacy. 2013;**9**(2):124-140

[88] Ribeiro DC, Costa S, Guilhermino L. A framework to assess the vulnerability of estuarine systems for use in ecological risk assessment. Ocean and Coastal Management. 2016;**119**:267-277

[89] Gonçalves A, Marques J. LTER Minho estuary [internet]. In LTER-ESTUARIES-Portugal. Avaliable from: https://data.lter-europe.net/deims/site/lter_eu_pt_011 [Accessed: 2019-02-14]

[90] Zacarias NG, DaSilva AJ. Tide propagation in the Minho River estuary (Portugal). In: 43rd Estuarine & Coastal Sciences Association – International Symposium. 2008

[91] Delgado A, Taveira-Pinto F, Silva R. Hydrodynamic and morphodynamic preliminar simulation of river Minho estuary. In: 6ª Jornadas de Hidráulica, Recursos Hídricos e Ambiente. 2011. pp. 113-126

[92] Portela LI. Morphological evolution, hydrodynamics and sediment dynamics of the Minho River estuary. In: 3o Seminário Sobre Gestão de Bacias Hidrográficas "Os Estuários." 2011. p. 6

[93] Zacarias NG. Influência da batimetria e do caudal fluvial na propagação da maré no estuário do rio Minho. Technical report, Universidade de Évora; 2007. p. 81

[94] Delgado A. Caracterização hidrodinâmica e sedimentar do estuário do rio Minho. [Thesis]. Universidade do Porto; 2011

[95] Sousa MC, Vaz N, Alvarez I, Gomez-Gesteira M, Dias JM. Modeling the Minho River plume intrusion into the rias Baixas (NW Iberian Peninsula). Continental Shelf Research. 2014;**85**: 30-41

[96] Iglesias I, Avilez-Valente P, Bio A, Bastos L. Numerical modelling simulations of the Minho estuary hydrodynamics. In: IX Simpósio Ibérico sobre a Bacia Hidrográfica do Rio Minho. 2018. p. 25

[97] Dias JA, Taborda R. Tidal gauge data in deducing secular trends of relative sea level and crustal movements in Portugal. Journal of Coastal Research. 1992;**8**(3):655-659

[98] Antunes C. Análise da evolução do nível médio do mar em Cascais. Entregável 1.1.6.a of the Estudo do litoral na área de intervenção da APA, I. P. /ARH do Tejo 2013; Available from: https://sniambgeoviewer.apambiente.pt/Geodocs/geoportaldocs/Politicas/Agua/Ordenamento/SistemasMonitorizacaoLitoral/E_1.1.6.a_Evoluao_NMM.pdf

[99] Araújo IB, Bos MS, Bastos LC, Cardoso MM. Analysing the 100year sea level record of Leixões, Portugal. Journal of Hydrology. 2013;**481**:76-84

Section 2

Coastal Dynamics and Ongoing Climate Processes in Coastal and Marine Environments

Chapter 3

Long-Term Changes in Sea Surface Temperature Off the Coast of Central California and Monterey Bay from 1920 to 2014: Are They Commensurate?

Laurence C. Breaker

Abstract

We examine to what extent the waters of Monterey Bay act independently of those along the central California coast. Sea surface temperatures (SSTs) from 1920 to 2014 from the central California coast and Monterey Bay were analyzed for long-term trends. To estimate the trends, singular spectrum analysis and empirical mode decomposition were employed. Between 1920 and 1940, long-term trends inside and outside Monterey Bay revealed rapidly increasing temperatures. After 1940 trends inside the bay indicate that temperatures increased from ~1950 for the next 40 years, peaking around 1990, and then decreased rapidly through 2013. Offshore, temperatures increased to the early 1960s, after which they decreased until 2014. El Nino episodes, the Pacific decadal oscillation (PDO), and increased coastal upwelling contribute to the long-term trends. Also, the impact of regime shifts associated with the PDO may be sustained for decades. Overall, the differences in the trends inside and outside Monterey Bay are significant only during summer where large-scale processes dominate offshore, and smaller-scale processes are important in and around the bay. Finally, our results suggest that waters inside the bay, although they co-vary with the waters further offshore, often appear to behave independently based on the long-term trends.

Keywords: long-term trends, nonlinear trends, singular spectrum analysis, ensemble empirical mode decomposition, central California coast, Monterey Bay, El Nino, Pacific decadal oscillation, coastal upwelling, regime shifts

1. Part I: background and preliminaries

1.1 Introduction

This study seeks to determine if the waters off the central California coast and Monterey Bay have warmed significantly during the 94-year period from 1920 to 2014 by examining sea surface temperatures (SSTs) inside the bay and outside the bay off the central California coast. The period of observation was terminated in December 2013 due to the unexpected arrival of a massive temperature anomaly off the coast of central California called the "Blob" in early 2014 [1]. We also seek to determine if warming inside the bay differs from warming outside the bay. This comparison is motivated, in part, by the following question. Is the bay merely an extension of the waters further offshore along the central California coast and thus expected to have similar physical properties, or are there local processes within or near the bay that significantly alter those properties? One of the arguments that favors the similarity of these waters is due to the relatively large entrance of the bay compared to its longest internal dimension (approximately 36 vs. 42 km). Thus, the waters offshore have wide and direct access to the bay. Also, residence times in the bay are not overly long, on the order of 7 days [2]. Thus, there is relatively less time for local waters that enter the bay to be modified before they exit. Early work in the greater Monterey Bay area by Skogsberg [3] and Skogsberg and Phelps [4], for example, tended to favor similarity. However, several of the more recent process-oriented studies that have been conducted on finer temporal and spatial scales in or near the bay would favor dissimilarity [5, 6]. In the simplest case, if we were to find significant differences in the long-term trends of temperature inside and outside the bay, we might favor dissimilarity.

To make these determinations, we calculate the long-term trends in SST in both domains. The results depend to a certain degree on how we distinguish between "coastal" waters and waters further offshore and on the methods that are used to estimate these trends. In this study we examine not only the long-term trends but also the nature of the processes that most likely have contributed to the trends.

The coastal ocean off central California and Monterey Bay is strongly affected by the process of coastal upwelling. According to Garcia-Reyes and Largier [7], coastal upwelling off central and northern California occurs from April through June, followed by relaxation of the upwelling-favorable winds from July through September. During the winter months from December through February, extratropical storms occur which contribute to cooler surface temperatures through wind mixing and the transfer of sensible and latent heat. Other times of the year tend to be transitional.

Based on the work of Mendelssohn and Schwing [8] and Garcia-Reyes and Largier [9], coastal upwelling off central California has been shown to have increased over the past several decades. According to Bakun [10] and Snyder et al. [11], coastal upwelling may be expected to increase in the future due to climate change or long-term climate variability.

Coastal upwelling is not the only physical process in the California Current System (CCS) that has a significant impact on SST. Ocean fronts, Ekman pumping, eddies, and squirts and jets all affect temperature. With respect to ocean fronts, the front that separates upwelled waters near the coast from oceanic waters further offshore is a primary example [12]. Positive wind stress curl off the central California coast during the spring and summer leads to Ekman pumping or offshore upwelling due to Ekman divergence. This is an important process that brings colder waters to the surface away from the influence of a coastal boundary. Both cyclonic and anticyclonic eddies are found in the CCS. Cyclonic eddies promote upwelling and are often found south of capes along the California coast. A major anticyclonic eddy is occasionally observed just west of Monterey Bay [13]. Jets and squirts occur off central and northern California characterized by patterns of vigorous circulation, readily observed in satellite images of SST [14].

During the spring and summer, upwelled waters are found inside Monterey Bay that often originate at Pt. Ano Nuevo (**Figure 1**), are advected down the coast, and then enter the bay forming a cyclonic pattern of circulation that has frequently been observed [15, 16]. In addition to upwelled waters that are advected into the bay from further offshore, local upwelling occurs in northern Monterey Bay due to the diurnal sea breeze that is well developed during the summer [5].

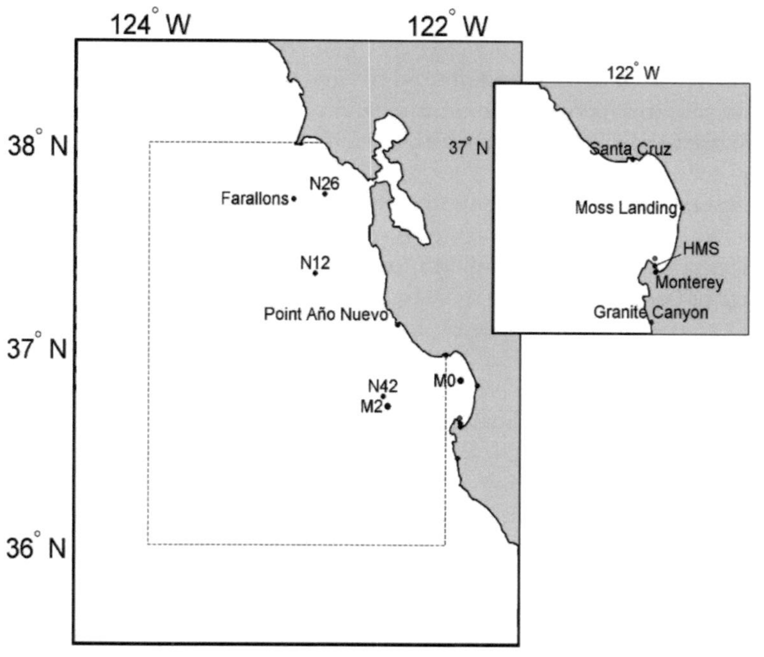

Figure 1.
*A map of the study areas including the offshore domain shown by the dotted box and the inshore domain which is Monterey Bay. The location where the data used to evaluate the SST from HMS is shown in the inset by the gray dot just above (north) of the Hopkins Marine Station (HMS). N26, N12, and N42 refer to National Data Buoy Center (NDBC) buoy locations (***Figure 4***).*

Now we turn to a review of the previous work that is directly relevant to this study. Inside Monterey Bay, SSTs from the Hopkins Marine Station at the southern end of the bay (**Figure 1**) have been examined on several occasions for possible long-term trends [17–19]. The record begins in 1919 and extends up to the present time. In each case, warming rates of approximately +0.01°C/year were found, based on linear, least-squares fits to the data.

Barry et al. [17] found that the annual maximum in SST increased more rapidly than the annual minimum. Further, they observed that changes in the flora and fauna of the intertidal zone in Monterey Bay coincided with well-documented secular warming along the US West Coast. In addition, they indicated that climate-related faunal changes in California's rocky intertidal community were related to long-term changes in the coastal environment based on the SST data from Hopkins and inferred that these changes were due to climate warming.

Sagarin et al. [18] indicated that temperature changes in Monterey Bay tended to be monotonic when the Hopkins record was subdivided into selected time periods between 1920 and 1995. They also concluded that climate warming, based on the Hopkins record and other sources, was responsible for range-related shifts in the intertidal communities along the California coast.

In a previous study, Breaker [19] examined the record at Hopkins for the period from 1920 to 2001 and also estimated the long-term linear trend. He found the slope to be +0.011°C/year, consistent with the values obtained by Barry et al. and Sagarin et al. He further found that the trend was statistically significant at the 95% level of confidence. Breaker examined the seasonal effects on long-term warming and found that warming in July exceeded the yearly rate by ~7%, whereas warming in

January was less than the yearly rate by almost 19%, consistent with the results of Barry et al. [17]. In the same study, Breaker also estimated the relative importance of the various processes that contributed to variability in the data and found that the annual cycle and its first harmonic, El Nino warming episodes, the Pacific decadal oscillation (PDO), and the long-term trend were all significant sources of variability.

The present study departs significantly from the previous study in a number of ways. First, one of the questions from the earlier work that was left unanswered was to what extent did the results from Monterey Bay reflect what was happening on larger scales outside the bay. Second, the record is now almost 15 years longer than the record used by Breaker [19], and significant cooling has occurred during this period. Third, in the previous study, only the long-term *linear* trend was calculated, and, since then, it has become clear that using a linear basis to model long-term changes in the data is a poor choice. Finally, we apply singular spectrum analysis to the data together with empirical mode decomposition to obtain independent estimates of the long-term trends. The methods are complementary and provide a basis for evaluating their quality.

Off the coast of central California between 35°N and 39°N, Garcia-Reyes and Largier [9] examined long-term trends in SST and several related parameters from the existing network of NDBC buoys for the period from 1982 to 2008. During the period of observation, the upwelling-favorable alongshore winds increased and SST decreased, consistent with increased coastal upwelling. Not only was the upwelling stronger, the length of the upwelling season was found to have increased, starting earlier in the spring and ending later in the fall. Relevant to this study, the observed trend in SST was found to be strongest off central California. Further, the observed cooling was limited to coastal waters over the shelf, where upwelling is the dominant process during the spring and summer. Because the record was only 27 years long, it was not possible to determine if the apparent trends were related to decadal or longer-term periodic behavior.

A larger view of the CCS was taken by Field et al. [20] who examined long-term warming from extensive records of SST off California and from other regions of the North Pacific. Much of their data spanned the twentieth century although it did not resolve the region of coastal upwelling off California. For locations where SSTs were examined in the North Pacific, the Pacific decadal oscillation (PDO) index accounted for a significant fraction of the observed variability. Also, near-surface temperature variations throughout the CCS were found to be similar between regions. Their data indicate that from 1900 on, strong negative SST anomalies prevailed during the period leading up to the early 1920s. They resulted from higher atmospheric pressure off California that intensified flow in the California Current and upwelling during this period. During the twentieth century, data from the CCS revealed warming trends with slopes ranging from +0.007°C/year to +0.010°C/year, similar to the warming observed in Monterey Bay over approximately the same period. Also, the negative anomalies in the early years of the twentieth century have contributed significantly to the positive trends that have been observed in records that are long enough to include this period.

Finally, we return to the original question posed in the title of this chapter. Although the word "commensurate" can refer to many different measures of similarity, we take it to mean to what extent do the waters inside Monterey Bay co-vary with the waters outside the bay with respect to their physical properties. During the winter the influence of the poleward flowing Davidson Current dominates the behavior of waters inside and outside the bay, and thus SSTs are generally similar [13]. During the summer when coastal upwelling is a dominant process, the relationship between the waters inside and outside the bay becomes more complicated because smaller-scale processes in and around Monterey Bay become important. Thus, the answer to this question may depend to a significant degree on the season, with higher co-variability to be expected during the winter than during the summer. Our subsequent analyses will shed more light on this question.

1.2 Data sources

1.2.1 Data sources inside the bay

The primary source of SST data inside the bay comes from the Hopkins Marine Station (HMS) located at the southern end of Monterey Bay in Pacific Grove (**Figure 1**). The data have been acquired nominally at 08:00 am PST on a daily basis since January 20, 1919. Because of several issues that have affected data quality over its duration, Breaker et al. [21] reconstructed the record by making time-of-day adjustments for varying data collection times, removed gaps, improved data resolution consistency, and, finally, reconstructed one entire year (1940) that was missing from the record based on regression with daily SSTs from the Farallon Islands. For the purposes of this study, we have extracted the 95+-year period from January 1920 through May 2015 although most of the results we present are limited to the period from 1920 through 2013 to avoid the confounding effects of the major temperature anomaly that arrived in early 2014. The daily observations were then averaged to obtain monthly mean values. Because these data have been acquired at a single location, they represent point observations. Daily observations of SST adjacent to HMS were also collected from August 1, 2006, through January 31, 2007, to ascertain the representativeness of the data from Hopkins per se and are discussed in Section 3.

1.2.2 Data outside the bay

The primary source of data for the 2° × 2° region shown in **Figure 1** along the coast of central California is the International Comprehensive Ocean-Atmosphere Data Set (ICOADS). These data were provided by the NOAA/OAR/ESRL PSD, Boulder, Colorado, USA, from their web site at http://www.esrl.noaa.gov/psd/. Monthly averaged SSTs were acquired for the region from 36 to 38°N, and from 122 to 124°W, for the period from January 1920 through May of 2015. Again, most of the results we present are limited to the period that ends in December 2013. These monthly data represent spatially averaged values and as a result are inherently different than the point observations we employ inside the bay.

Figure 2 shows the number of observations per month in the 2 × 2 degree study area starting January 1920. Overall, the number of observations remains below 1500 per month until 2004 when a significant increase occurs. The inset shows the period from 1920 through 1970 where the numbers of observations are far smaller.

Plots of the monthly averaged SST data from the Hopkins Marine Station, inside Monterey Bay, and from ICOADS, adjacent to the central California coast, outside the bay, are shown in **Figure 3**. Superimposed on the data are smoothed versions shown in red obtained using a *LOWESS* smoothing function. The method is nonparametric and performs robust, locally weighted regression. It fits linear or quadratic basis functions to the data at the center of neighborhoods where the radius of each neighborhood contains a specific percentage of the data points. This type of smoothing function does not lose degrees of freedom at the ends of the record and introduces minimal distortion at these locations. The fraction of data in each neighborhood, and thus the smoothness, is determined by two parameters, the degree of the local polynomial basis function (linear or quadratic), λ, and the level

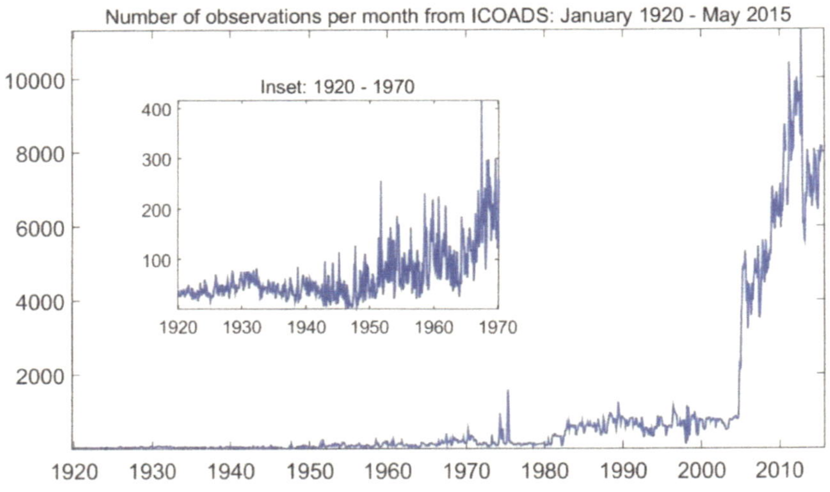

Figure 2.
The number of observations per month from ICOADS for the offshore domain are plotted by month since 1920. The inset shows the period between 1920 and 1970 where the number of observations per month is relatively small.

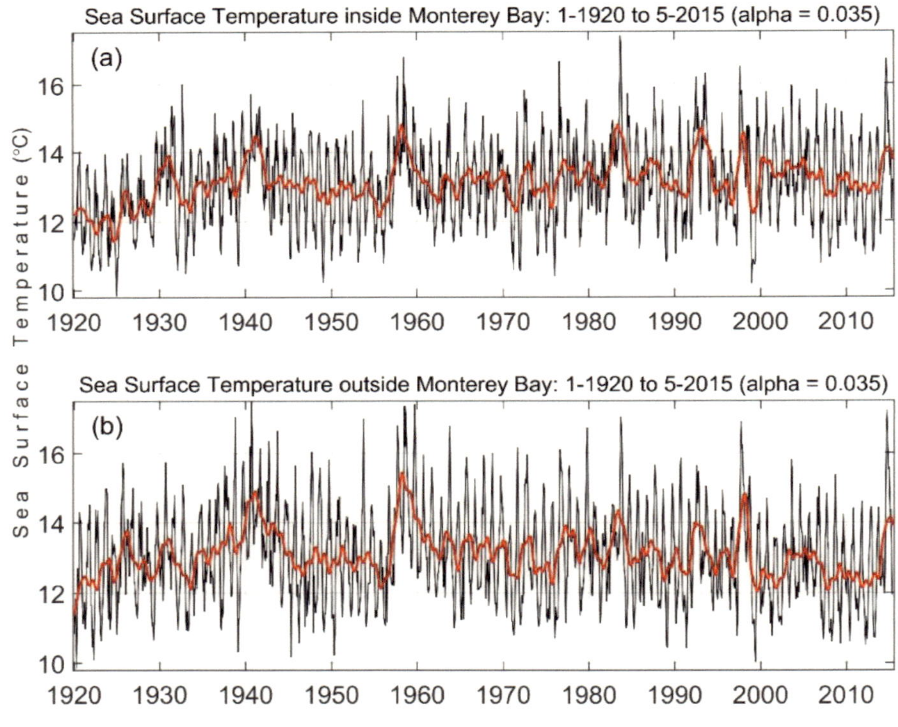

Figure 3.
Sea surface temperature from HMS inside Monterey Bay (a), and from ICOADS, off the coast of central California, outside the bay (b). The red curves provide smoothed versions of the data where important features such as major El Nino warning episodes can be more easily identified.

of smoothing, α. In the present study, the degree of the local polynomial has been set to quadratic. The smoothing parameter is specified by the choice of α, where $0 \leq \alpha \leq 1$. For α equal to 0, no smoothing occurs, and for α equal to 1, we obtain essentially a straight line. The goal is to choose α large enough to obtain as much smoothness as possible without distorting underlying patterns in the data [22].

At the level of detail shown in **Figure 3**, the relative importance of the annual cycle is readily apparent. It is also possible to identify major El Nino warming episodes that occurred in 1931–1932, 1940–1941, 1958–1959, 1982–1983, 1986, 1992–1993, and 1997–1998. In this study we usually refer to El Nino warming events or simply El Ninos rather than ENSO which has a broader meaning that includes La Nina cooling events as well. Off central and northern California, it is the warming phase of the El Nino phenomenon that stands out and not the cooling phase.

Also, as mentioned earlier, there was a major increase in SST that is readily apparent in the data inside and outside the bay during 2014. We show it here but do not include it in our calculations. According to **Figure 3**, the increase in SST based on the unsmoothed data during 2014 clearly exceeds 3°C, somewhat larger than the reported value of 2.5°C [1].

1.3 Data representativeness

Inside the bay, the SST data are acquired at the shoreline in a somewhat sheltered area next to the Hopkins Marine Station. Next to the coast where the circulation is expected to be weaker, residence times could be longer, leading to increased temperatures, locally. To examine this question which has arisen before, we compared the SST data from HMS with SSTs acquired at an adjacent offshore location. Specifically, daily SSTs at HMS were compared with daily SSTs acquired just offshore, slightly north of HMS in southern Monterey Bay (**Figure 1**). The data were acquired over a period of almost 6 months from August 4, 2007, through January 31, 2008. This data was acquired specifically to evaluate the daily observations of SST that have been collected at HMS (M. Denny, personal communication). The two data sets are almost identical with a mean difference of less than 0.1°C over the 6-month period of observation. Based on an earlier comparison of SST data between HMS and Santa Cruz, located at the north end of Monterey Bay (**Figure 1**), Breaker [19] concluded that the data from HMS was generally representative of SST data collected elsewhere in the bay, consistent with the results presented here.

The offshore data used in this study have been acquired over a spatial domain that is regional in scale (**Figure 1**). The number of observations per month has increased significantly since 1920 (**Figure 2**). This, in turn, changes the uncertainty in estimating the monthly means over the length of record. This problem is not unique in climate research. The globally averaged record of surface air temperature used extensively in climate studies suffers from the same problem.

Spatial biases in the SST observations acquired over the offshore domain lead to biases in the monthly mean values that are calculated. This arises for at least two reasons. First, the observations that enter into the monthly means are not uniformly spaced over the region since they come from a variety of sources. Second, even if the observations were acquired uniformly, the underlying ocean itself is not thermally homogeneous. The second factor becomes particularly apparent during the spring and summer when coastal upwelling occurs. During this period SSTs may increase by 5°C or more between the coast and the offshore boundary of the study domain. These biases are significantly reduced between November and February during the Davidson Current period when the waters tend to become more isothermal. Overall, spatial biases are to be expected, they vary with the season, and they will influence the trends we calculate.

To examine the spatial bias problem we have drawn from Garcia-Reyes and Largier [9] who examined SST and other data for the period from 1982 through 2008 from eleven NDBC Environmental Data Buoys along the California coast. Three of those buoys (N26, N12, and N42) are located in our offshore study area (**Figure 1**). They calculated the linear trends in SST for the upwelling season which they defined as the period from March through July for all buoys. We have taken our offshore data, with the buoy data specifically removed, and calculated the linear trend for the same time period and upwelling season (**Figure 4**). For buoys N26, N12, and N42, Garcia-Reyes and Largier obtained slopes of −0.032, −0.034, and −0.021°C/year, respectively, with a mean of −0.0290°C/year, taken over the three buoys. Based on our data, we obtained a slope of −0.024°C/year, slightly less than, but still relatively close to, the value obtained by Garcia-Reyes and Largier. If the non-buoy data were randomly spaced over the domain, we would find this result encouraging in that it suggests a certain degree of spatial homogeneity in the data. If, however, the non-buoy data were concentrated closer to the coast, as we suspect, then all we have shown is that the buoy and non-buoy data tend to be consistent.

1.4 Long-term trends

1.4.1 Background

Although trends may appear to be intuitively obvious, precise definitions are hard to find. According to Fuller [23], the term "trend" only acquires meaning when a specific procedure is used for its estimation. For time series, trends have been labeled as deterministic when the trend per se is modeled explicitly [24]. A trend implies non-stationarity of the first order, and so its removal may transform the data into a stationary process. Linear trends are often fitted to the data, but as we shall see in the present study, the long-term changes we observe are not linear. In this regard, two studies are relevant. Jevrejeva et al. [25] analyzed sea level data for long-term trends that were nonlinear in nature using a variation of singular spectrum analysis (SSA) called Monte Carlo SSA. This method provides error estimates associated with the trend as well. Breaker et al. [26] examined sea surface temperature data from the coast of Ecuador for long-term trends that were inherently nonlinear using ensemble

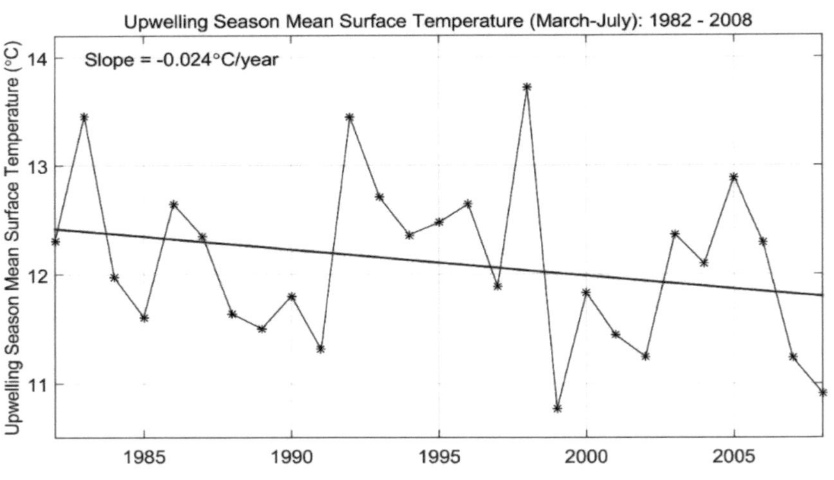

Figure 4.
Mean seasonal SST for the offshore domain for the upwelling period from March through July plotted on a yearly basis for 1982 through 2008. A least-squares linear fit to this data yields a slope of −0.024°C/year.

empirical mode decomposition. This method performed well in extracting the long-term trends regardless of the degree to which the data tended to be nonlinear.

In our experience, when the data are trend-dominant, different methods of estimating the trend often yield similar results, but when the trend is not a dominant feature in the data, then different methods may yield significantly different results. One common definition of the trend is that of a smoothly varying function whose first derivative does not change sign [27]. Although this definition may be intuitively appealing, in practice, it is often too restrictive. Although formal definitions of the trend are scarce, Wu et al. [28] recently proposed the following: "The trend is an intrinsically fitted monotonic function or a function in which there can be, at most, one extremum within a given data span." For rather extensive reviews of methods for estimating trends, see Esterby [27] and Alexandrov et al. [29].

1.4.2 Methods

To estimate long-term trends in the data, we use two methods of spectral decomposition, singular spectrum analysis (SSA) and ensemble empirical mode decomposition (EEMD). We find that SSA and EEMD are often complementary in the trends they produce, and, when they are, it increases our confidence that meaningful trends have indeed been extracted from the data. Both methods decompose the data into a set of independent modes. To this extent, the methods are similar. However, the methodologies per se are entirely different. We give a brief introduction to each method here and the appropriate references.

Singular spectrum analysis (SSA) decomposes a time series into a set of independent modes, similar in many respects to principal component analysis. The method is well-suited for analyzing data that are nonstationary and/or nonlinear due to the adaptive nature of the basis functions employed. A lagged covariance matrix (a Toeplitz matrix, in this case) is formed from the time series that is decomposed into eigenvalues, eigenvectors, and principal components. Reconstructed components can be calculated from the eigenvectors and principal components that represent partial time series whose sum over all modes reproduces the original time series. These components, modes, or eigentriples as they are variously called are often amenable to physical interpretation. The number of modes that is selected is called the window length and determines the resolution of the decomposition. The mode or modes that correspond to the trend can often be selected by inspection when the trend contains a significant portion of the variance, but when the trend is relatively weak, the problem becomes more difficult. Alexandrov [30] provides a new approach for extracting the trend when it is not a dominant feature in the data. In this study, the trend, although not the most dominant feature in the data, is robust enough that we do not resort to the method of Alexandrov in order to isolate it. Finally, we note that because the method of decomposition in SSA is global in nature, problems can arise at the boundaries of the data. For further details concerning SSA, see [31–35].

Empirical mode decomposition (EMD) is a method of decomposing a time series into a sequence of empirically orthogonal intrinsic mode function (IMF) components and a residual. In EMD, the number of modes is determined by the data, whereas in SSA, the number of modes is a free parameter that must be specified by the user. Like SSA, EMD is data adaptive and well-suited for the analysis of nonstationary and nonlinear time series. The IMF components are often physically meaningful because the characteristic scales in each case are determined by the data itself. As in SSA, selected modes may require grouping in order to extract a physical basis. Each IMF represents a mode of oscillation with time-dependent amplitude and frequencies that lie within a specific band of frequencies, the center of which defines the mean period of that mode. The process of extracting the individual modes or essential scales from the data is called sifting and is performed many times to produce a single IMF. In practice, extracting the trend may require that not only the residual but one or several of the highest adjacent modes be grouped in order to reconstruct it.

One problem in the application of EMD in the past was that mode mixing often occurred when a time series included intermittently occurring signals with widely separated time scales. To address this problem, EMD now includes a noise-assisted component in its calculation. Wu and Huang [36] developed the technique that is now called "ensemble EMD," or EEMD, which defines the true IMF as the mean of an ensemble of IMFs and, in the process, preserves the physical uniqueness of the decomposition. An ensemble member consists of the signal plus white noise of finite amplitude. The magnitude of the white noise that should be added is given by the ratio of the standard deviation of the first IMF to the standard deviation of the data itself and is called N_{std}. Although this is the recommended value, in many cases, if slightly different values above or below the recommended value are used, the resulting IMF patterns remain stable although the patterns can differ slightly. Typically, the number of realizations or ensemble size is several hundred in order to generate the ensemble. The ensemble size that we have used in this study is 300. Thus, in EEMD there are two free parameters that must be specified, the level of white noise to be added and the ensemble size. The problem of end effects is discussed in Wu and Huang [36]. In the latest version, the IMFs themselves are extended after they are calculated which helps to reduce problems at the boundaries. The problem is further reduced in the noise-assisted version, i.e., in EEMD, because the slopes of the IMFs tend to be more uniformly distributed in the ensemble. For more information concerning EMD and EEMD, see [36–40].

2. Part II: analyses, results, and discussion

2.1 Analyses

2.1.1 Application of SSA and EEMD

In applying SSA to the data inside and outside the bay, the decomposition was accomplished using a window length, L, of 240 months inside the bay, and L = 316 months in the offshore domain (outside the bay). Although we started with L = 240 months in the second case, due to the occasionally encountered problem of mode mixing (the partial presence of one mode in an adjacent mode), we increased L in steps to its present value in order to suppress it. In SSA terminology this issue is referred to as the separability problem and is addressed in detail in Golyandina et al. [32]. **Figure 5** shows the first 12 reconstructed modes for L = 316 outside the bay. The first two modes (Ro1 and Ro2) correspond to the annual cycle, and modes 6 and 7 (Ro6 and Ro7) correspond to the first harmonic. Harmonic components in the data often appear as matched pairs in the reconstructed modes where the amplitudes and phases are the same or similar [41], making it easier to identify them. Finally, we note that mode 5 is trend-like in nature.

Inside the bay, the corresponding SSA decomposition indicates that modes 1 and 2 correspond to the annual cycle and modes 9 and 10 to its first harmonic (not shown). In this case, mode 3 is trend-like and reveals two maxima, one in the early 1940s and the second around 1990, reminiscent of the maxima that we associate with the Pacific decadal oscillation (PDO). At this point, we tentatively associate

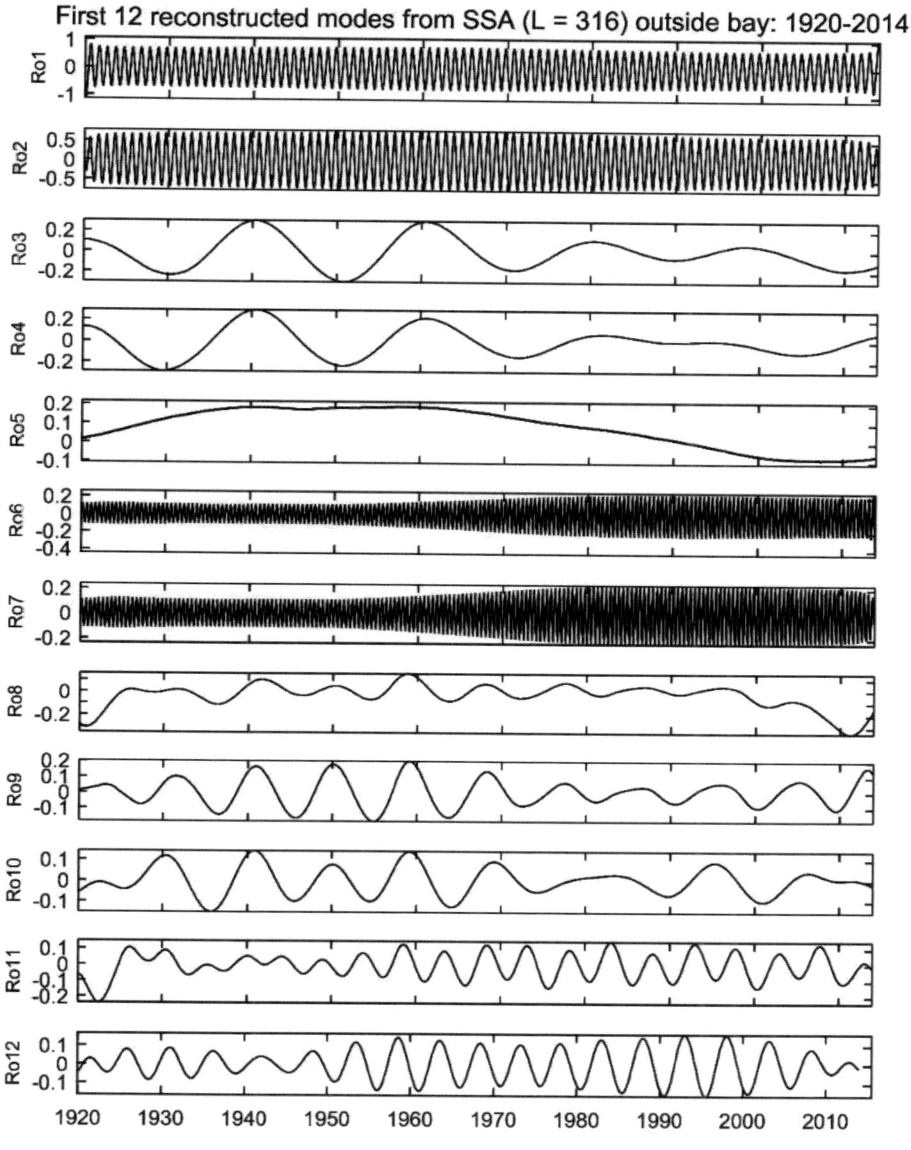

Figure 5.
The first 12 reconstructed modes are shown from the SSA decomposition of the data (L = 316) outside the bay. The first two modes show the annual cycle, and modes 6 and 7 show the first harmonic of the annual cycle. Reconstructed mode 5 shows the long-term trend.

this mode and mode 5 from the decomposition outside the bay with the underlying trends, awaiting the results from the EEMD decompositions.

In decomposing the data inside the bay based on EEMD, the noise parameter, N_{std}, was determined to be 0.33 and the ensemble size was 300. The EEMD decomposition produced 10 modes (imf1 – imf10) where imf10 is often referred to as the residual. In a similar decomposition of the data outside the bay, the value for N_{std} was 0.31 with the same ensemble size, and, again, 10 modes were produced. The results from inside the bay (black) for all 10 modes are shown in **Figure 6**, together with the three highest modes (imf8, imf9, imf10—blue) from outside the Bay. The results obtained inside and outside the Bay are generally similar except for the highest three modes, and thus the reason why they are included in the same figure. Also, it is these modes that most likely contribute to the long-term trends.

Next, we calculate and plot the variances for each mode inside and outside the bay from the EEMD decompositions (**Figure 7**). If we can identify the oceanic processes in accordance with the modal decompositions, then we may be able to estimate their relative importance. The estimated center frequencies for each mode (i.e., band) and location are given in **Table 1**. The center frequencies are not fixed because they are determined by the data and so vary slightly between locations [42]. The variances tend to be similar overall between locations but there are several notable differences. The variances associated with the first modes are similar

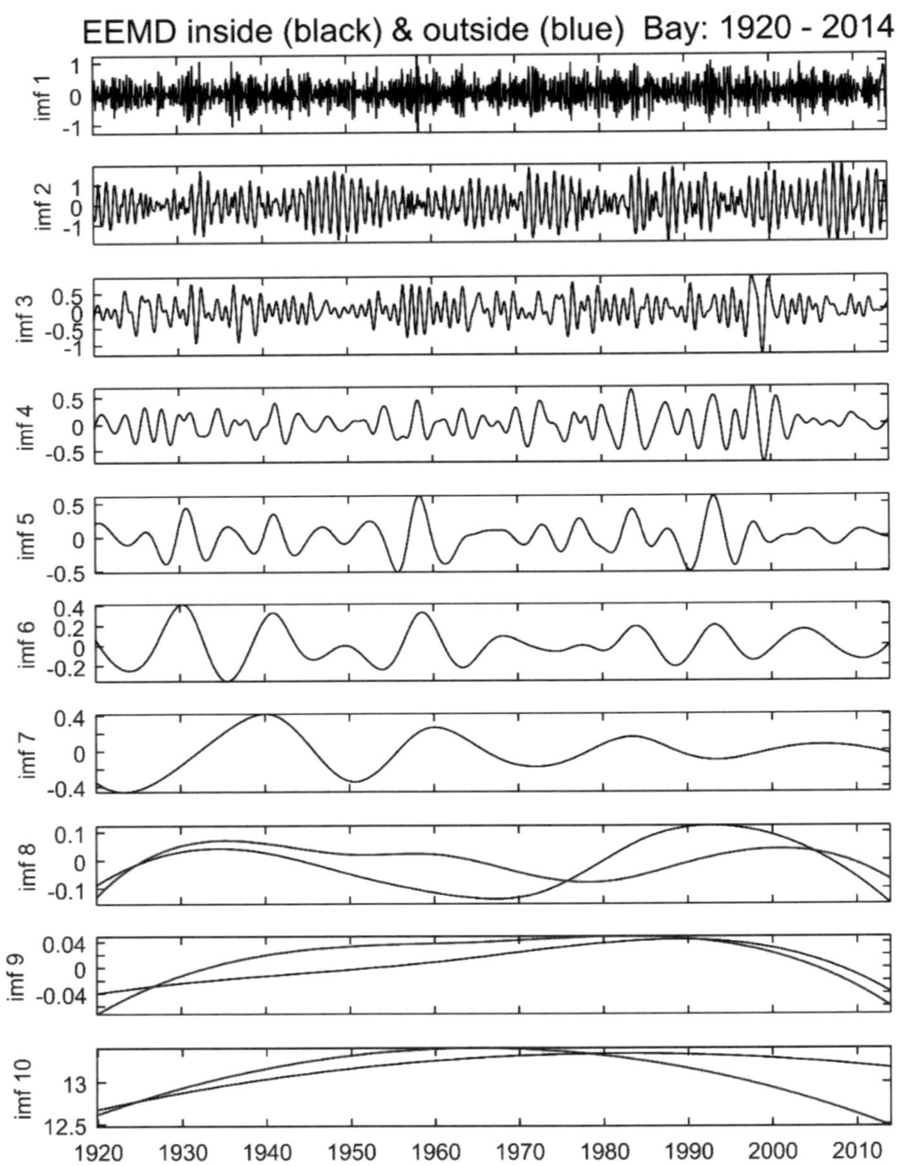

Figure 6.
EEMD decompositions of SST from inside and outside Monterey Bay each generated 10 modes. All 10 modes are shown for inside the bay (black) and the highest 3 modes (imf8 – imf10) are shown for outside the bay (blue) as well. See the text for details.

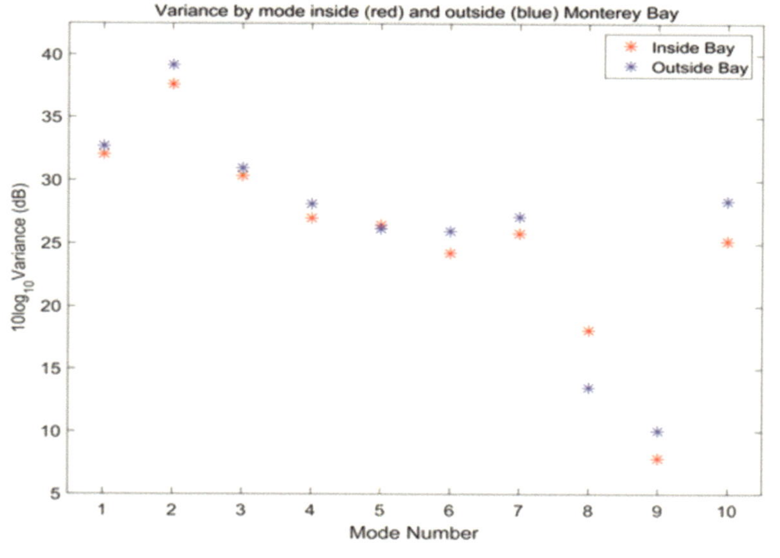

Figure 7.
Variances by mode obtained from the EEMD decompositions for observations acquired inside the bay (red) and outside the bay (blue).

IMF mode number	Inside bay	Outside bay	Comments
1	3.13 (3.8 months)	3.95 (3.0 months)	
2	1.0 (1.0 year)	1.0 (1.0 year)	
3	0.55 (1.8 years)	0.55 (1.8 years)	Depends on smoothing
4	0.31 (3.2 years)	0.328 (3.0 years)	
5	0.17 (5.9 years)	0.16 (6.3 years)	
6	0.082 (12.2 years)	0.06 (16.7 years)	
7	0.047 (21.3 years)	0.036 (27.8 years)	
8	0.012 (83.3 years)	0.012 (83.3 years)	
9	—	—	Insufficient data
10	—	—	Insufficient data

Table 1.
Approximate center frequencies in cycles per year and the corresponding periods in months or years (in parentheses) for each mode from the EEMD decompositions.

and represent approximately 15% of the total variances. The primary periods are 3 months outside and 3.8 months inside the bay with most of the variability distributed over a band that extends from roughly 2– 6 months. A significant portion of the variability in this range should be related to coastal upwelling and upwelling-related processes such as offshore Ekman transport and the evolution of ocean fronts, eddies, jets, and squirts [43]. The second modes correspond primarily to the annual cycle and in both cases represent over 50% of the total variance. Modes 3 and 4 are transitional in nature in that they generally fall between the periods we associate with annual and interannual variability (**Figure 6**). By mode 5, interannual variability is dominant with the major El Ninos of 1957–1958, 1982–1983, 1992–1993, and 1997–1998 making their appearances at both locations. Modes 6–8 show a gradual transition from what are distinctly El Nino events (mode 6) to a signature that we associate with the Pacific decadal oscillation (PDO) in mode 8, in each case. The PDO index [44] characteristically has a maximum circa 1940 and a second maximum circa 1990 that are separated by a period of approximately 50 years (**Figure 12**). Of particular note, the variance associated with the PDO in mode 8 is far greater inside the bay than it is outside the bay (by at least 6 decibels which corresponds to a factor of 2). We discuss this curious result in more detail in the final section of the text. Mode 9 in both cases carries very little variance but could be related to the long-term trend. The residual long-term trends are shown in mode 10. Finally, the variance associated with the residual long-term trend outside the bay far exceeds that of the trend inside bay as shown (**Figure 6**).

2.1.2 Trend identification

The highest and the next two highest modes from EEMD inside and outside the bay are shown in **Figure 8a** and **b**. In more detail, the highest mode (imf10), the sum of modes 9 and 10 (imf9 + imf10), and the sum of modes 8, 9, and 10 (imf8 + imf9 + imf10) are shown inside (**Figure 8a**) and outside (**Figure 8b**) the bay. First, we note that the results of adding modes 9 and 10 together (i.e., imf8 + imf9—red) makes little difference, as expected, based on the small variance carried by mode 9. It is not until we include mode 8 (i.e., imf8 + imf9 + imf10—green) that we introduce more structure into the trends. We note that the added structure is primarily due to the inclusion of the PDO in our tentative definition of the long-term trend.

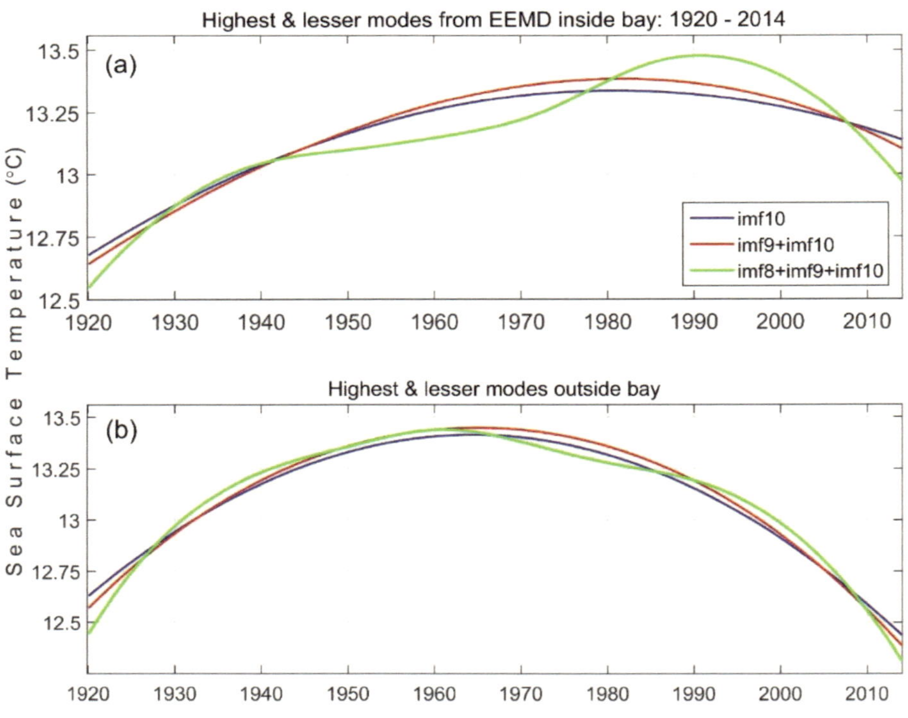

Figure 8.
In the upper panel (a), the highest mode (imf10—blue), the sum of the highest and next highest modes (imf9 + imf10—red), and the sum of highest, next highest, and third from the highest modes (imf8 + imf9 + imf10—green) are shown from the EEMD decomposition of the data from outside the bay. In the lower panel (b), the same combinations of modes from the EEMD decomposition of the data from inside the bay are shown.

The SSA results, however, provide a slightly different story. In deciding which mode or combination of modes corresponds to the long-term trend, the similarity between the tentatively identified SSA trends and the EEMD-derived trends obtained by combining modes 8, 9, and 10 becomes apparent. The trends from SSA and EEMD inside the bay are shown in **Figure 9a** and those trends outside the bay in **Figure 9b**. A high degree of similarity between SSA and EEMD is quite apparent inside the bay but to a lesser extent outside the bay.

Differences in the trends outside the bay can be explained, at least in part, by two factors. First, it was necessary to employ a larger value for the window length in the SSA decomposition outside the bay (L = 316 vs. 240 months) because of mode mixing that resulted in greater smoothing and thus a reduction in amplitude. The second problem relates to end effects. In our experience, SSA does not perform as well as EEMD when it comes to resolving structure near the beginning and end of the modal time histories. According to Mann [45], the problem of how to smooth data with trends near boundaries is particularly problematic. As stated earlier, SSA can exhibit problems at the boundaries of the reconstructed modes, but the boundary issues associated with EEMD, in our experience, are far smaller. Also, because the methods we have used are completely independent, there is no reason to expect that the combinations of modes which we choose to call the trend in one case should be the same or even similar in the other.

In summary, we favor the trends obtained from EEMD primarily because the problem of mode mixing that arose in the SSA decomposition outside the bay led to an unrealistic reduction in the amplitude of the reconstructed mode. Secondly, the end effects from EEMD are expected to be smaller than they are from SSA, and that has been our experience. It is important to note, however, that SSA has been most helpful in guiding our selection of the modes from EEMD to include or group in arriving at our definition of the long-term trends.

2.1.3 The trends per se

Based on the EEMD results, the trends inside and outside the bay are shown together in **Figure 9c**. They are clearly nonlinear and would be poorly approximated using a linear basis. During the 1920s the trends in SST are similar, but by 1930, the trends start to diverge with SSTs increasing more rapidly offshore than inside the bay. This pattern of divergence continues until about 1960 when SSTs inside the bay start to increase more rapidly and SSTs offshore start to decrease. Opposing trends in SST since 1960 continue for approximately the next 30 years. By about 1990, however, the trend inside the bay changes from increasing to decreasing temperatures, consistent with strongly decreasing temperatures offshore. Without the arrival of the warm anomaly in early 2014, both trends would most likely have continued to reflect decreasing temperatures up through at least 2014 and perhaps for several years longer.

2.1.4 Uncertainty

Uncertainty in climate science has been referred to as a "monster," a "wicked problem," or "the 800 pound gorilla who is sitting in the next chair" [46]. Mathematically, estimating uncertainty is considered to be an ill-posed problem because different methods of estimating this quantity often produce different results.

In our work, the question of uncertainty in estimating the trends shown in **Figure 10** naturally arises. As stated by Moore et al. [47] regarding nonlinear trends, we expect the confidence interval of a nonlinear trend to be significantly smaller than it would be for a least-squares fit to the same data, since the data are

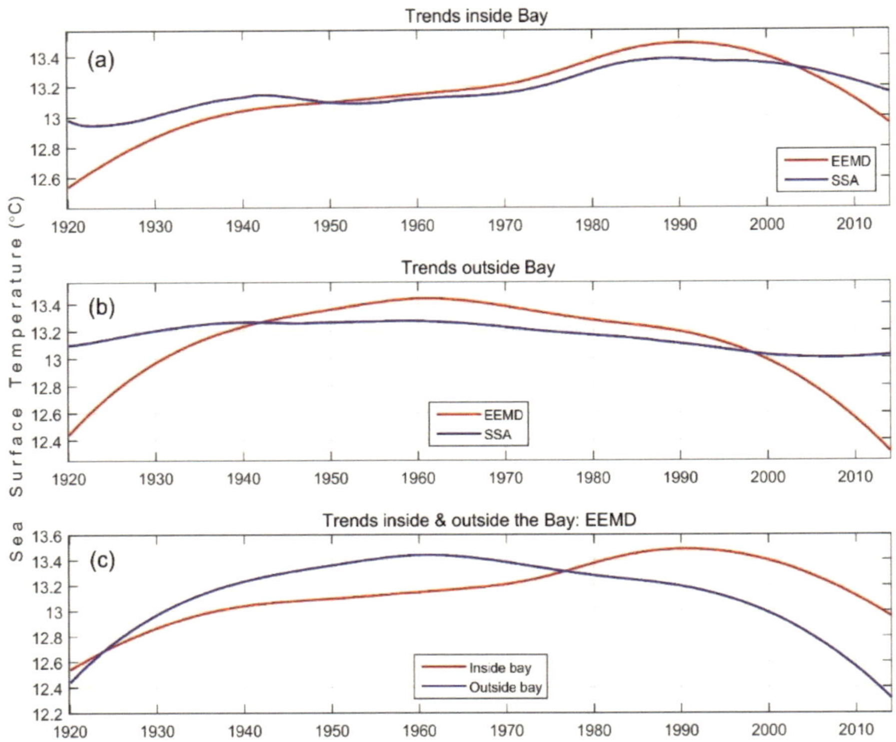

Figure 9.
The trends obtained from SSA (blue) and EEMD (red) of the data from inside the bay are shown in (a). The trends obtained from the data outside the bay using the same color convention are shown in (b). (c) shows the trends inside (red) and outside (blue) the bay together based on the results of EEMD.

not forced to fit any specified basis function. However, they acknowledge that it is not necessarily a simple problem to estimate uncertainty in nonlinear cases. In our case, since we have two independent estimates of the trends, the spread between them does provide at least a rudimentary estimate of the uncertainty if we assume that the true trend lies between the two estimated trends. Of course, we could also use a weighted average to reflect our greater confidence in the results from EEMD, but how to determine such weights is anything but clear. To proceed with equal weighting, one could go even further and assume that the two curves in each case correspond to ±1 standard deviation about the mean of a Gaussian distribution and, from that point, calculate the 95% confidence intervals. This approach is used in small sample theory in statistics when no other information is available (personal communication, Prof. David S. Crosby). This approach has the distinct advantage of using both estimates of the trend to estimate a confidence interval that would be associated with the expected value of the two trends obtained by calculating their mean.

We have proceeded to calculate confidence intervals for the trends inside and outside the bay. The results are shown in **Figure 10a** and **b**. There are several ways these intervals can be estimated. In this case, the differences between the trends and the mean values between them serve as a proxy for the standard deviations. We can calculate a global standard deviation for each record and use that estimate to obtain the 95% confidence intervals following the usual assumptions and tables given in any standard text on statistics. However, this approach yields confidence intervals that are constant over the entire record. A better approach in our view is to consider the proxy values locally and to calculate the 95% confidence intervals separately for each time point. This has the distinct advantage of emphasizing the increased uncertainty that arises at the end of each record at the price of indicating no uncertainty where the records intersect, a feature that is unrealistic. We also obtain the expected result that the uncertainty increases or decreases where the differences between the two records increase or decrease. Finally, we return to the question raised in Section 1.3 concerning the differences between the point observations acquired inside the bay, and the area-averaged observations acquired offshore. Although the uncertainties we have obtained address this question to some degree, simulations from an ocean model would be better-suited to address this question in detail.

2.2 Ocean processes and time scales

2.2.1 Seasonal variability

The modal patterns we have already observed are inherently smooth because they contain only the highest modes from the various EEMD decompositions. To examine the underlying processes that contribute to these patterns in more detail, we have stratified the data from inside and outside the bay by season where we define the summer season as the mean of June, July, and August and the winter season as the mean of November, December, and January. We have smoothed the results slightly using LOWESS for added clarity.

In **Figure 11a**, inside the bay (red) during summer, the patterns are strongly positive and are almost identical for the first two decades indicating that the process or

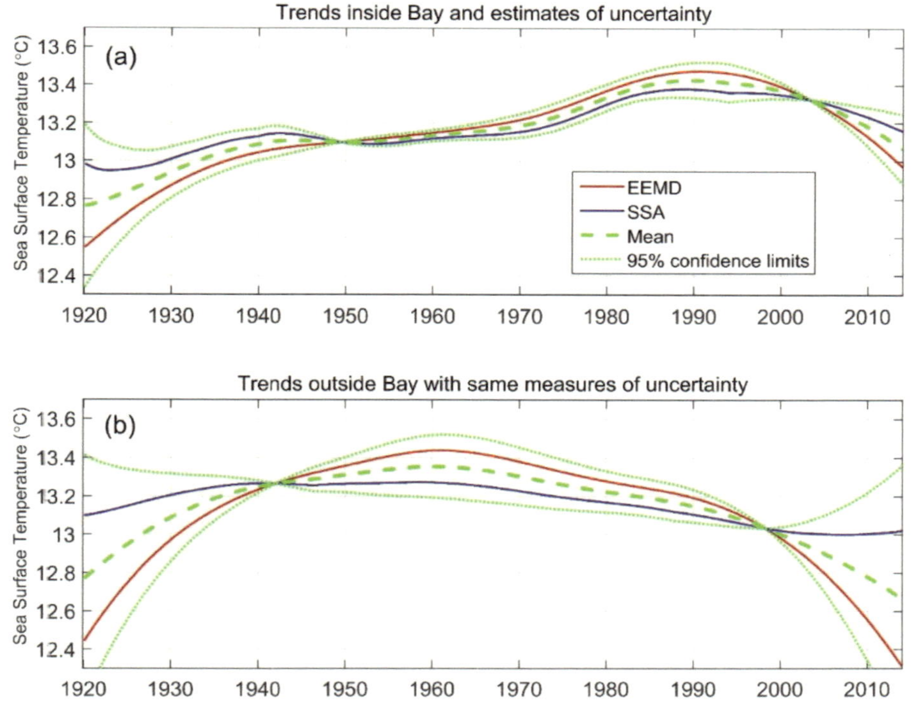

Figure 10.
The upper panel (a) shows the SSA- and EEMD-derived trends inside Monterey Bay together with one measure of uncertainty. The green dashed line shows the estimated true trend based on the mean of the SSA and EEMD trends. The green dotted lines show the 95% confidence intervals about the mean. In the lower panel (b), the original trends, the estimated mean value, and the 95% confidence intervals are again shown but, in this case, for the waters outside the bay.

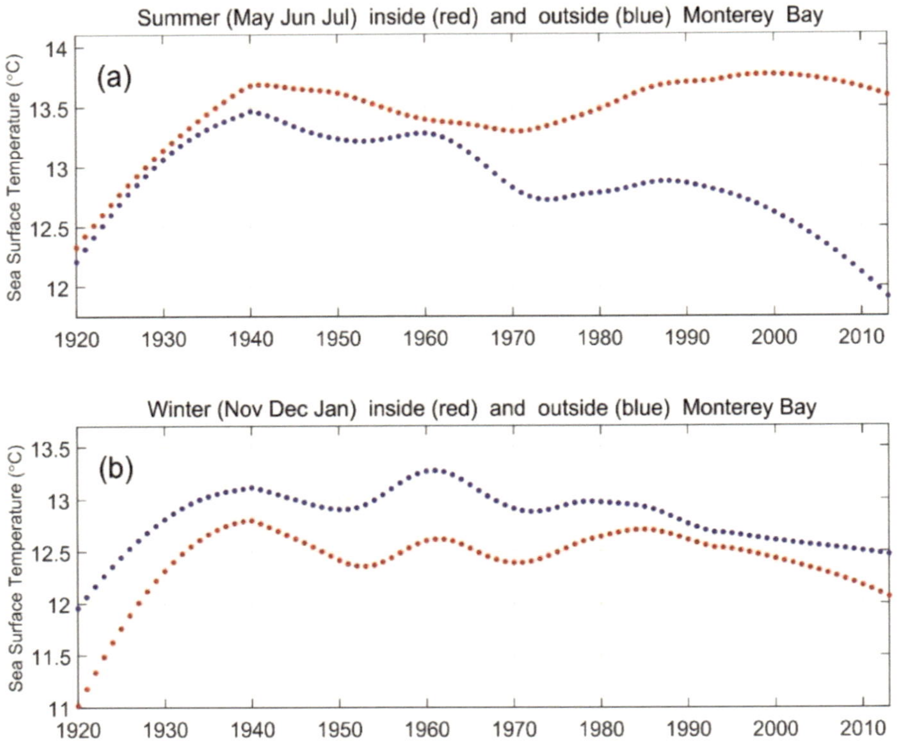

Figure 11.
In both panels, the data have been stratified by season where summer corresponds to the average of May, June, and July and winter corresponds to the average of November, December, and January. In the upper panel (a), the data for summer inside (red) and outside (blue) the bay are shown. In the lower panel (b), the data for winter inside (red) and outside (blue) the bay are shown. LOWESS smoothing has been applied in each case with the degree of smoothing (alpha) set equal to 0.35 (see text for details). Finally, the points have not been connected to emphasize the fact that they represent seasonal values, and so the curves that are plotted are not continuous functions.

processes responsible for contributing to the rapid warming included both domains. The rates of increase in temperature during this period approach 5°C/100 years! After 1940, these trend-like patterns moderate and gradually diverge, and by 2014 they have diverged to a point where bay waters are almost 1.5°C warmer than waters outside the bay. These patterns convey a more detailed picture of the variability that could not be observed in the original trends.

Continuing, inside the bay, although SSTs increased by over 1°C by 1940, between 1940 and the early 1970s, they actually decreased. The increase now occurs mainly between the early 1970s up to ~2000, after which they decrease until 2014. Outside the bay, SSTs actually decrease, albeit with significant variations along the way, until the end of the record (December 2013). Although the overriding pattern is downward, after the early 1970s, SSTs increase slightly until the late 1980s, after which they decrease rapidly, consistent with our previous observations.

Four points of note are as follows: (1) the patterns of rapidly increasing SST during the first two decades inside and outside the bay were almost identical, indicating that the warming process was a truly dominant phenomenon during that period; (2) SSTs actually decrease from 1940 to the early 1970s inside the bay; (3) cooling, for the most part, occurred outside the bay starting ~1940 (vs. ~1960) and continued until the end of the record; and (4) surface temperatures inside the bay were significantly higher than outside the bay during the summer, especially over the last 20 years or so.

In **Figure 11b**, during the winter, the patterns inside (red) and outside (blue) the bay are, to a high degree, similar, even with the higher resolution. However, although the patterns are similar, the temperatures outside the bay reflect SSTs that are roughly 0.2–1°C higher during winter. In more detail, maxima in temperature occur at both locations circa 1940 and during the early 1960s. These maxima co-occur with El Nino warming episodes with the addition of the co-occurrence of the first major maximum in the PDO index in or about 1940 (**Figure 12**).

What is common in all cases is the very rapid rates of warming during the first 20 years of each record and the slightly less rapid rates of cooling during the last 20 years or so. During each period the influence of the PDO is virtually unmistakable. Also, during the early years of the past century, coastal upwelling intensified due to an intensification of the subtropical high pressure system which overlies the central North Pacific during summer [20]. However, by the early 1920s, this intensification had apparently abated leading to weaker winds along the coast and reduced coastal upwelling. Weaker coastal upwelling is, of course, consistent with increasing SSTs along the coast which is what we have observed.

2.2.2 Interannual and interdecadal variabilities

As we have stated, based on the long-term trends inside and outside Monterey Bay, temperatures increased rapidly from 1920 through at least 1940 inside the bay. Outside the bay, temperatures continued to increase more gradually up to approximately 1960. A major El Nino occurred in 1940–1941 (**Figure 6**), and the Pacific decadal oscillation index (http://research.jsiao.washington.edu/pdo/PDO) had one of its two primary maxima during the past century circa 1940 (**Figure 12**) and, together, almost certainly contributed to these trends during this period. From the late 1940s to the late 1960s, trends inside the bay show little change, but by 1970 temperatures increase significantly up to the early 1990s, consistent with the massive El Nino episode of 1982–1983 plus a second maximum in the PDO circa 1990 (**Figure 12**). Inside the bay, by the mid-1990s, SSTs finally started to decrease, and they decreased steadily until 2014, again consistent with the strongly decreasing amplitude of the PDO during this period.

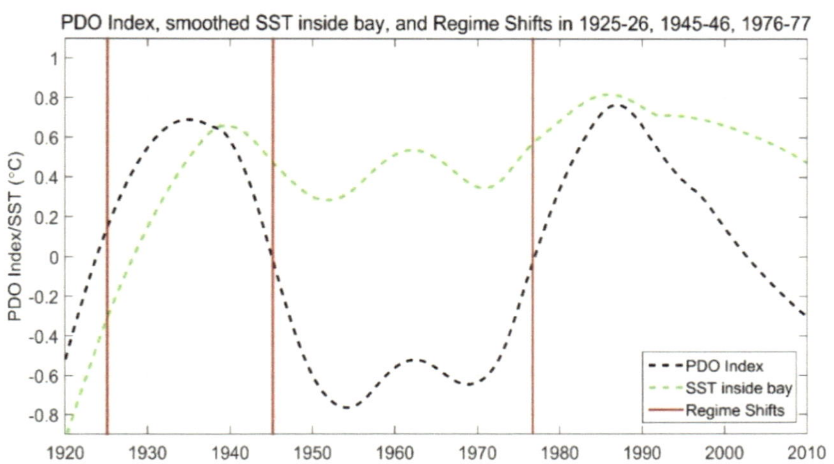

Figure 12.
A smoothed version of the Pacific decadal oscillation index (°C) is shown (dashed curve—black) together with a smoothed version of the SST record from inside the bay (dashed curve—green). Superimposed on these plots are three regime shifts that occurred during the past century (vertical red lines).

After 1960, trends outside the bay show a pattern that is in opposition to the trend inside the bay with SSTs decreasing steadily up to the early 1990s, consistent with the cooling trends observed by Garcia-Reyes and Largier [9]. By the mid-1990s, the rate of cooling increased significantly through 2013 and is likely due to the added influence of the PDO, a period where the amplitude of the PDO index likewise decreased rapidly.

2.2.3 The PDO and the impact of regime shifts

Two processes that have received considerable attention are El Nino warming events and the PDO because of their expected importance in contributing to the warming (cooling) process. The time scales of El Nino warming events off central California vary, but they last for at least 6 months and can persist for periods of up to 2 years. The El Nino episodes in 1940–1941, 1958–1959, and 1982–1983 are examples of events that lasted for almost 2 years. Although these events are transient, they may leave an imprint that is imbedded in memory of the ocean for decades [48].

The time scales of the long-term oscillations that characterize the PDO are 20–30 years [49]. However, the changes in physical properties associated with the change in phase of the PDO are far shorter. These changes, i.e., regime shifts, have time scales that are of the order of 6 months [50]. Since 1920, three major regime shifts have occurred: 1925–1926, 1945–1946, and 1976–1977. These events have each been associated with a phase change in the PDO.

Figure 12 shows a smoothed version of the PDO index together with a smoothed version of the SST data from inside the bay. We note that the events in 1925–1926 and 1976–1977 occur when the slope of the PDO index is positive, leading to changes in temperature along the California coast that were positive, whereas the event in 1945–1946 occurred when the slope of the PDO was negative [50]. In this case, the event led to a decrease in long-term temperature along the coast of California. It is important to note that although the signs of these changes along the coast of California were as indicated above, in other parts of the North Pacific, basin changes of opposite sign in SST were observed with respect to 1976–1977 event [51, 52].

We now take a closer look at these events. Because they are buried in the day-to-day natural variability of the data, they are difficult to detect and localize. Breaker [50] used cumulative sums to identify and localize regime shifts that have occurred since the early 1920s. Regime shifts produce a characteristic pattern in the cumulative sum that allows us to estimate when the event occurred, its duration, and its midpoint, t_o.

Once a regime shift has occurred, are the changes that occur short-lived or are they sustained over longer periods? To address this question, we employed a method called the expanding mean [51]. We start at the midpoint of these events and calculate the mean values proceeding forward and backward in time and then compare the results after we have proceeded in each direction for several years or longer.

Near t_o, the mean values vary widely because only a few observations enter into the calculation, but after 2–3 years, the plots generally start converging to mean values that become relatively stable. The influence of El Nino episodes was avoided wherever possible. These plots were generated for each event.

The expanding mean plot for the 1925–1926 regime shift showed that stable mean differences first appeared at 1.5–2 years out from t_o, with values in the range of +0.3°C estimated between years 2 and 3.5 in the forward direction. The differences remain consistently positive out to year 6, but a major El Nino event in 1930 precluded a more extensive comparison. The expanding mean plots for the 1945–1946 regime shift revealed that a decrease in SST did occur along the central California coast. Here, the differences are greater after 2–3 years and were consistently negative out to 10 years. The El Nino of 1940–1941 notwithstanding, we estimate a mean decrease of at least −0.4°C. For the 1976–1977 event, from 3 to 6 years forward from t_0, the differences were consistently positive with values that ranged from about +0.1°C to almost +0.5°C. Overall, after the first ~2.5 years, positive differences between the two curves were sustained out to almost 10 years, with increases, on average, that were of the order of +0.2°C.

These results suggest that small but sustained changes in the mean value of temperature occurred following the 1925–1926, 1945–1946, and 1976–1977 regime shifts. These changes appear to have been sustained for periods of up to a decade and perhaps for the entire period between regime shifts. As a result, these events, although brief in nature, may well contribute to the long-term trends that we have sought to identify.

2.3 Discussion

The long-term trends obtained in the foregoing analyses and presented earlier were obtained using singular spectrum analysis (SSA) and ensemble empirical mode decomposition (EEMD). The results tend to be similar within each domain although significant differences arose. Based on our experience in this study and others, we favor the results obtained using EEMD for the reasons we discuss in Section 4.3.2. However, SSA provided valuable guidance in helping us decide on a suitable definition for the long-term trend.

Monthly averaged SSTs from January 1920 through December 2013 for two adjacent areas, the central California coast and Monterey Bay, have been examined for long-term trends based on the results from EEMD. These trends show that from 1920 to 1940, temperatures increased rapidly in both domains. After 1940, the trends inside and outside the bay are basically different. Inside the bay the trends indicate that temperatures tended to increase from about 1950 through 1990, while outside the bay, they decreased continuously from about 1960 through 2013. Temperatures inside the bay also decreased after 1990 until the arrival of a major thermal anomaly in 2014. Of particular note is the period during the 1970s and 1980s where trends inside the bay indicate that temperatures increased at rates higher than at any other time prior to 1940. In early 2014, a major temperature anomaly referred to as the "Blob" arrived along the central California coast that lasted for almost 2 years [1], and made it virtually impossible to continue the analysis beyond that point since our primary goal was to compare long-term trends inside Monterey Bay and outside the bay off the open coast.

Based on our nonseasonal results from EEMD, although SSTs inside the bay have decreased rapidly since the early 1990s, overall, if we consider the entire record, they have increased from about 12.6°C to almost 13.0°C over the 94 years from 1920 to 2014. Outside the bay SSTs may not have changed much over the entire 94-year period. They reached a maximum of slightly over 13.4°C circa 1960 with values of about 12.4°C in 1920 and a value of about 12.3°C at the end of 2013. For comparison, although we do not show the results from linear analyses that were performed, they indicate that the linear trend inside the bay is positive and statistically significant, whereas the trend outside the bay is negative and is not statistically significant.

To take a closer look at the data, a higher resolution analysis based on smoothing was conducted. The data were decomposed into two seasons, summer (MJJ) and winter (NDJ). Based on these decompositions (**Figure 11**), significant warming inside the bay since the 1970s has occurred only during the summer. Thus, there is most likely a connection to coastal upwelling, albeit indirect. Outside the bay, SSTs decreased continuously during this period. According to Garcia-Reyes and Largier [9], SSTs along the central California coast have decreased since the early 1980s due to increased coastal upwelling which is directly related to increased upwelling-favorable winds. Our results are consistent with theirs where the data overlap, and so it is likely that the same processes have been at work back to at least the 1960s, and perhaps a decade or two earlier, off central California.

According to Bakun [10], the increase in coastal upwelling over the past several decades is related to increased heating in inland California and the surrounding area that has intensified the thermal low pressure system in the Southwestern United States. This intensified low pressure system together with the subtropical high pressure system off the coast of California has increased the pressure gradient between the two regions. This increase in the onshore-offshore pressure gradient in turn produces stronger winds along the California coast and thus more intense coastal upwelling. The increase in inland heating may be due to sustained climate warming or simply to long-term climate variability.

Why increased warming occurred inside the bay until ~1990, while cooling occurred continuously off the coast of central California since about 1960, is a curious fact if we assume that the behavior of bay waters is directly related to what happens further offshore. One possibility is that as coastal upwelling increased off the coast, more intense upwelling took place in the upwelling center off Point Ano Nuevo, located approximately 40 km north of Monterey Bay (**Figure 1**), whose waters are advected down the coast toward the bay. This flow is most likely a manifestation of the coastal jet which typically flows in close proximity to the coast [53]. This flow bifurcates when it reaches the bay and part flows into the bay and part flows offshore [15, 16]. If the down-coast flow increased as upwelling increased in the Ano Nuevo upwelling center, then the process of bifurcation or the location where it occurred could have been altered. Such a change might well have resulted in less upwelled water entering the bay and more going elsewhere. As a result, with less cold, upwelled water entering the bay, SSTs would have increased. In this case, the waters inside and outside the bay may still co-vary, but this co-variation could take the form of an inverse relationship rather than a direct one.

Continuing, why then did temperatures inside the bay start decreasing in the early 1990s, becoming more in line with what had been taking place outside the bay since the early 1960s? It is possible that what occurred during the 1960s inside the bay may have resulted in a basically unstable flow regime that could not be sustained indefinitely and so it eventually returned to its original state. Other possibilities exist, but this question presents an ideal opportunity to employ a hydrodynamic model of Monterey Bay and the surrounding area to address it.

To summarize the seasonal analyses, regardless of the season, there were very high rates of warming during the first 20 years of each record and only slightly less rapid rates of cooling after ~1990. The very high rates of increasing temperature are in agreement with Field et al. [20] who found that SSTs off southern California rose rapidly by the mid-1920s due to secular warming, consistent with the rapidly increasing SSTs we have observed off central California and Monterey Bay during the same period.

Before we leave the seasonal results, we point out that most of what has been said and will be said regarding the various analyses refers primarily to those results obtained from the EEMD modal decompositions, i.e., the long-term trends, and not from the seasonal analyses although these results have added additional insight at the shorter time scales.

In comparing the physical properties and flow along the central coast and Monterey Bay between November and February, i.e., during winter, waters inside and outside the bay are dominated by poleward flow associated with the Davidson Current [13]. As a result we expect that these waters tend to co-vary in unison and that the physical properties are similar. The results shown in **Figure 11b** are consistent with this view. However, during summer the situation is very different (**Figure 11a**), and so it is not surprising that differences in the trends occur for the reasons we have just described.

A number of processes contribute to the long-term trends in SST that we have observed. Interpreting these trends in terms of the oceanic processes that contribute to them is not necessarily straightforward. However, from the modal variances obtained from our EEMD decompositions, we can identify a number of these processes and, to some degree, estimate their relative importance. We start with coastal upwelling. It is one of the dominant processes along the California coast. The first mode from each EEMD decomposition contains variability with periods that range from less than 2 months to slightly greater than 6 months. In each case it accounts for approximately 15% of the total variance. According to our results and those of Garcia-Reyes and Largier [9], coastal upwelling has continued to intensify over at least the past several decades, in agreement with Bakun [10]. Offshore, SSTs have generally decreased since as early as 1940 based on our seasonal analyses. Thus, it is clear that coastal upwelling is a major contributor to lower SSTs offshore and may have contributed indirectly to higher SSTs inside Monterey Bay.

A number of upwelling-related processes also fall within a similar range of time scales as coastal upwelling. These include offshore Ekman transport, Ekman pumping, the spring transition, and squirts and jets. Although not entirely separate from coastal upwelling, the growth and decay of cyclonic and anticyclonic eddies and the evolution of ocean fronts are included in this range. In the last case, with regard to eddies, although their life cycles often exceed 6 months, much of the variability associated with them occurs on shorter time scales.

Although the annual cycle is the major contributor to the variances in SST inside and outside the bay, they have little or no apparent impact on the long-term trends. Conversely, El Nino warming episodes and the Pacific decadal oscillation are clearly important. It is difficult to estimate the variances associated with the El Nino and the PDO separately because they occupy at least four adjacent modes in our EEMD decompositions (**Figure 6**). According to the sequences, the El Nino episodes gradually coalesce into the PDO as we progress from mode 5 to mode 8 in each case. A similar coalescence takes place inside the bay for the same modes. Together, these modes account for only ~10% of the total variance, but the processes they represent make the greatest contributions to the long-term trends because they have the longest time scales. Finally, consistent with our observations, it is now well recognized that the PDO owes its existence in large measure to the ENSO phenomenon [54–56].

Although the variance of the PDO is relatively small in each case, its influence on the long-term behavior of the data is highly significant. Of particular note, our decompositions revealed that the variance of the PDO, although relatively small compared to the other modes, is far greater inside the bay than it is outside the bay. As a result, we would expect its impact on the long-term trends to be greater inside the bay. That this is the case is clearly shown in **Figure 9** where the maximum value of the long-term trend inside the bay occurs in the early 1990s and thus coincides with one of the two major maxima in the PDO index (**Figure 12**). Why the amplitude of the PDO is greater inside the bay may be due to its containment within a limited domain where its energy tends to accumulate more rapidly than it can dissipate. If we are correct, then this relationship may hold in other coastal regions where similar embayments are found.

Unlike El Nino warming events that are intermittent in nature, the influence of the PDO, although less apparent, may be both continuous and transient. Transient influences may occur when there are phase changes in the PDO. These phase changes often correspond to regime shifts. The time scale of these events is on the order of 6 months. Our results suggest that small but sustained changes in the mean value of temperature occurred following the 1925–1926, 1945–1946, and 1976–1977 regime shifts. These changes appear to be sustained over periods of up to a decade or longer. Also, regarding the 1976–1977 event, in two related studies, one off the coast of Hawai'i [48] in the central Pacific and the second off the coast of South Korea [49] in the western Pacific, it was found that SSTs decreased rather than increased as they had off central and southern California during the 1976–1977 event [47]. We conclude by stating that the influence of the PDO may be twofold where its total impact is composed of continuous and transient contributions.

Finally, the long-term trend in SST inside the bay departs significantly from the trend further offshore over the 94-year period from 1920 to 2014. However, the differences in these trends were far greater during the 30-year period between 1960 and 1990 which imply that significant changes in the local circulation in and around the bay must have occurred during that period. We make an important distinction, however, between the differences in the trends we have observed and the extent to which they reflect independent behavior. It is likely that during the 1970s and 1980s, for example, where the trends differ significantly, these waters still co-varied but the relationship between them was not direct but indirect as we have said before. Thus, upwelling off the coast could be increasing at roughly the same time that upwelled waters entering Monterey Bay are decreasing.

3. Conclusions

Although the results of Garcia-Reyes and Largier [9] indicate that coastal upwelling along the coast of California has intensified since the early 1980s, our results suggest that this cooling process has been at work since at least the early 1960s and possibly earlier.

Nonlinear methods were used throughout this study to estimate the long-term trends. In addition to fitting long-term variability in the data more realistically, nonlinear trends will, in most cases, have smaller confidence intervals than those associated with the corresponding linear trends. Also, of importance, the nonlinear trends often provide more insight into the processes that contribute to them.

The major thermal anomaly that occurred off the coast of central California in early 2014 made it virtually impossible to conduct a meaningful analysis of the long-term trends because of its magnitude and duration. Thus, this chapter serves as a period piece that spans a recent 94-year period during which we were able to observe and hopefully better understand the relationship between the waters of Monterey Bay and the waters further offshore along the central California coast.

We often find a sensitive interplay between resolving the trend and trying to interpret it. As the resolution is increased, new features often emerge that make it difficult to decide what part of the long-term variability contributes to the trend and what part does not. This problem was encountered in 2.1.2 of the present study.

Our results show that the amplitude of the Pacific decadal oscillation (PDO) is approximately twice as large inside Monterey Bay as it is outside the bay. This relationship most likely derives from the fact that Monterey Bay is partially enclosed. As a result, energy associated with the PDO that enters the bay may accumulate more rapidly than it can be dissipated, leading to higher amplitudes within the interior, overall. If we are correct, then we might expect to find a similar relationship in other coastal regions where embayments exist.

We conclude the warming that occurred during the first two decades of our data sets, i.e., from 1920 to 1940, was highly unusual. During this period SSTs increased at rates approaching 5°C/100 years. We know that SSTs off southern California also experienced a rapid increase in SST by the mid-1920s [20], and that during this same period, global SSTs likewise increased significantly [57].

Using spectral decompositions of the data, we were able to illustrate over a sequence of modes the process of El Nino episodes evolving or coalescing into the Pacific decadal oscillation, consistent with recent theoretical results that emphasize the contribution that El Ninos make to the formation of the PDO [54–56].

Although it would appear that the waters inside and outside Monterey Bay often act independently, this may not be the case. We do concur that these waters, to a large extent, co-vary. The point is that apparent independence and co-variation are not mutually exclusive terms. Although we have shown at least one way in which these terms can be reconciled, other possibilities exist.

In this study we examined three regime shifts that occurred in 1925–1926, in 1945–1946, and in 1976–1977 and found that small changes in the mean temperature, of order 0.5°C or less, accompanied these events based on the data from Monterey Bay. These regime shifts coincided in each case with a phase change in the PDO. Although these changes were relatively small, they were often sustained for periods of up to a decade or longer, and this tendency may be one of their defining characteristics.

The physical behavior of regime shifts is not completely understood. However, it is becoming apparent, based on limited observations, that some regions within the Pacific basin exhibit an increase in SST, while at others, a decrease is observed during the same event, in this case, during the 1976–1977 regime shift. The spatial distribution of these changes in sign around the North Pacific also raises the possibility that what we may be observing, if we connect the dots, is a standing wave pattern. Such a standing wave would most likely have a wavelength governed by the dimensions of the basin, motion governed by the phase of the Pacific decadal oscillation, and a period of approximately 20–30 years, based on the intervals between the events we have examined. Periods in this range are in close agreement with [58] who estimated the mean interval between events to be 23 years based on the historical PDO record that dates back to 1650.

Finally, why is the work presented here of value? One reason is that by obtaining a better understanding of the relationship between bay waters and those further offshore, we are better able to predict the behavior of one system when we only have information regarding the other. In the signal processing world, such a relationship would be referred to as a transfer function.

Acknowledgements

Steve Worley and Zaihua Ji from NCAR are thanked for providing the SST data from ICOADS. Dr. Mark Denny from the Hopkins Marine Station is thanked for providing SST data adjacent to the Hopkins Marine Station for the purpose of evaluating the Hopkins SST data. David Crosby, Professor Emeritus of Statistics at American University, is thanked for suggesting the method of estimating the uncertainty associated with the trends obtained using SSA and EEMD. Finally, one anonymous reviewer is thanked for providing a number of helpful comments and suggestions for improvement that have been incorporated in the final version of the text.

Author details

Laurence C. Breaker[1,2]

1 Moss Landing Marine Laboratories, Moss Landing, CA, United States of America

2 University of Delaware, Newark, DE, United States of America

*Address all correspondence to: laurence.breaker@gmail.com

References

[1] Bond NA, Cronin MF, Freeland H, Mantua N. Causes and impacts of the 2014 warm anomaly in the NE Pacific. Geophysical Research Letters. 2015;**42**:3414-3420. DOI: 10.1002/2015GL063306

[2] Lipphardt BL, Small D, Kirwan AD Jr, Wiggins S, Ide K, Grosch CE, et al. Synoptic Lagrangian maps: Application to surface transport in Monterey Bay. Journal of Marine Research. 2006;**64**:221-247. Available from: http://www.journalofmarineresearch.org

[3] Skogsberg T. Hydrography of Monterey Bay, California: Thermal conditions, 1929-1933. Transactions of the American Philosophical Society. 1936;**29**:152. Available from: https://www.amphilsoc.org/publications

[4] Skogsberg T, Phelps A. Hydrography of Monterey Bay, California: Thermal conditions, part II, 1934-1937. Transactions of the American Philosophical Society. 1946;**90**:350-386. Available from: https://www.amphilsoc.org/publications

[5] Woodson CB et al. Local diurnal upwelling driven by sea breezes in northern Monterey Bay. Continental Shelf Research. 2007;**27**:2289-2302. DOI: 10.1016/j.csr.2007.05.014

[6] Graham NE, Largier JL. Upwelling shadows as nearshore retention sites: The example of northern Monterey Bay. Continental Shelf Research. 1997;**17**:509-532. DOI: 10.1029/2009JC005623

[7] Garcia-Reyes M, Largier J. Seasonality of coastal upwelling off central and northern California: New insights including temporal and spatial variability. Journal of Geophysical Research. 2012;**117**:C12013, 15 pages. DOI: 10.1029/2011JC007629

[8] Mendelssohn R, Schwing FB. Common and uncommon trends in SST and wind stress in the California and Peru-Chile current systems. Progress in Oceanography. 2002;**53**:141-162. Available from: https://www.infona.pl/resource/bwmeta1.element.elsevier-027a36a5-dd40-39a3-a8ab-9e2456f06956/tab/citations

[9] Garcia-Reyes M, Largier J. Observations of increased wind-driven coastal upwelling off central California. Journal of Geophysical Research. 2010, 2010;**115**:C04011, 8 pages. DOI: 10.1029/2009JC005576

[10] Bakun A. Global climate change and intensification of coastal ocean upwelling. Science. 1990;**247**:198-201. DOI: 10.1126/science.247.4939.198

[11] Snyder MA, Sloan LC, Diffenbaugh NS, Bell JL. Future climate change and upwelling in the California Current. Geophysical Research Letters. 2003;**30**:1823, CLM, 8 pages. DOI: 10.1029/2003GL017647

[12] Mooers CNK, Flagg CN, Boicourt WC. Prograde and retrograde fronts. In: Oceanic Fronts in Coastal Processes. Springer-Verlag; 1978. pp. 43-58. DOI: 10.1007/978-3-642-66987-3_6

[13] Breaker LC, Broenkow WW. The circulation of Monterey Bay and related processes. Oceanography and Marine Biology. Annual Review. 1994, 1994;**32**:1-64. ISSN 0078-3218

[14] Lasker R. Food chains and fisheries: An assessment after 20 years. In: Towards a Theory on Biological-Physical Interactions in the World Ocean. Kluwer; 1988. pp. 173-182. Available from: https://link.springer.com/chapter/10.1007/978-94-009-3023-0_9

[15] Rosenfeld LK, Schwing FB, Garfield N, Tracy DE. Bifurcated flow from an upwelling center: A cold water source for Monterey Bay. Continental Shelf Research. 1994;**14**:931-964. DOI: 10.1016/0278-4343(94)90058-2

[16] Tracy DE. Source of cold water in Monterey Bay observed by AVHRR satellite imagery [thesis]. Monterey, California: Naval Postgraduate School; 1990. p. 126

[17] Barry JP, Baxter CH, Sagarin RD, Gilman SE. Climate-related, long term faunal changes in a California rocky intertidal community. Science. 1995;**267**:672-675. Available from: https://www.ncbi.nlm.nih.gov/pubmed/17745845

[18] Sagarin RD, Barry JP, Gilman SE, Baxter CH. Climate-related change in an intertidal community over short and long time scales. Ecological Monographs. 1999;**69**:465-490. Available from: http://www.bioone.org/doi/pdf/10.3398/042.007.0120

[19] Breaker LC. What's happening in Monterey Bay on seasonal to interdecadal time scales? Continental Shelf Research. 2005;**25**:1159-1193. DOI: 10.1016/j.csr.2005.01.003

[20] Field DB, Baumgartner TR, Charles C, Ferriera-Bartrina V, Ohman MD. Planktonic foraminifera of the California Current reflect twentieth century warming. Science. 2006;**311**:63-66. DOI: 10.1126/science.1116220

[21] Breaker LC, Broenkow WW, Denny MW. Reconstructing an 83-Year Time Series of Daily Sea Surface Temperature at Pacific Grove, California. Scripps Institution of Oceanography Library. Paper 7; 2006. Available from: http://repositories.cdlib.org/sio/lib/7

[22] Cleveland WS. Robust, locally weighted regression and smoothing scatterplots. Journal of the American Statistical Association. 1979;**74**:829-836. Available from: http://links.jstor.org/sici?sici=0162-1459%28197912%2974%3A368%3C829%3ARLWRAS%3E2.0.CO%3B2-L

[23] Fuller WA. Introduction to Statistical Time Series. Wiley Series on Probability and Statistics; 1996. pp. 475-476. Available from: https://www.amazon.com/Introduction-Statistical-Time-Wayne-Fuller/dp/0471552399

[24] Hamilton JD. Time Series Analysis. Princeton: Princeton University Press; 1994, First chapter. Available from: https://press.princeton.edu/links.html

[25] Jevrejeva S, Grinsted A, Moore JC, Holgate S. Nonlinear trends and multiyear cycles in sea level records. Journal of Geophysical Research. 2006;**3**:C09012. DOI: 10.1029/2005JC003229

[26] Breaker LC, Loor HR, Carroll D. Trends in sea surface temperature off the coast of Ecuador and the major processes that contribute to them. Journal of Marine Systems. 2016;**164**:151-164. DOI: 10.1016/j.jmarsys.2016.09.002

[27] Esterby SR. Review of methods for the detection and estimation of trends with emphasis on water quality applications. Hydrological Processes. 1996;**10**:127-149. DOI: 10.1002/(SICI)1099-1085(199602)10:2%3C127::AID-HYP354%3E3.0.CO;2-8

[28] Wu Z, Huang NE, Long SR, Peng C-K. On the trend, detrending, and variability of nonlinear and nonstationary time series. Proceedings of the National Academy of Sciences. 2007;**104**:14889-14894. DOI: 10.1073/pnas.0701020104

[29] Alexandrov T, Bianconcini S, Dagum EB, Maas P, McElroy T. A

review of some modern approaches to the problem of trend extraction. Econometric Reviews. 2012;**31**:593-624. Available from: http://scholar.google.de/citations?user=NUjVCEwAAAAJ&hl=de

[30] Alexandrov T. A method of trend extraction using singular spectrum analysis. Revstat—Statistical Journal. 2009;7:1-22. Available from: https://arxiv.org/abs/0804.3367

[31] Elsner JB, Tsonis AA. Singular Spectrum Analysis: A New Tool in Time Series Analysis. Plenum Press; 1999. pp. 51-65. DOI: 10.1007/978-3-642-34913-3

[32] Golyandina N, Nekrutkin V, Zhigljavsky A. Analysis of Time Series Structure: SSA and Related Techniques. Chapman & Hall/CRC; 2001. pp. 44-53. ISBN 1-58488- 194-1

[33] Ghil M et al. Advanced spectral methods for climatic time series. Reviews of Geophysics. 2002;**40**:1-41. DOI: 10.1029/2000RG000092

[34] Golyandina N, Zhigljavsky A. Singular Spectrum Analysis for Time Series. Springer; 2013. pp. 37-39. DOI: 10.100/978-3-642-34913-3

[35] Golyandina N, Korobeynikov A, Zhigljavsky A. Singular Spectrum Analysis with R. Springer, eBook; 2018. DOI: 10.1007/978-3-662-57380-8

[36] Wu Z, Huang NE. Ensemble empirical mode decomposition: A noise-assisted data analysis method. Advances in Adaptive Data Analysis. 2009;**1**:1-41. DOI: 10.1142/S1793536909000047

[37] Huang NE et al. The empirical mode decomposition and the Hilbert spectrum for nonlinear and non-stationary time series analysis. Proceedings of the Royal Society of London, Series A. 1998;**454**:903-995. DOI: 10.1098/rspa.1998.0193

[38] Huang NE. Introduction to Hilbert-Huang transform and some recent developments. In: The Hilbert-Huang Transform in Engineering. Taylor & Francis; 2005. pp. 1-24. ISBN-13: 978-0-8493-3422-1

[39] Huang NE. Introduction to the Hilbert-Huang transform and its related mathematical problems. In: The Hilbert-Huang Transform and its Applications. World Scientific; 2005. pp. 1-26. ISBN 981-256-376-8

[40] Huang NE, Wu Z. A review on Hilbert-Huang transform: Method and its applications to geophysical studies. Reviews of Geophysics. 2008;**46**:1-23. DOI: 10.1029/2007RG000228

[41] Vautard R, Yiou P, Ghil M. Singular spectrum analysis: A toolkit for short, noisy, chaotic signals. Physica D. 1992;**58**:95-126. DOI: 10.1016/0167-2789(92)90103-T

[42] Flandrin P, Goncalves P, Rilling G. EMD equivalent filter banks from interpretation to applications. In: The Hilbert-Huang Transform and its Applications. World Scientific; 2005. pp. 57-74. Available from: http://perso.ens-lyon.fr/patrick.flandrin/HHT05.pdf

[43] Checkley DM, Barth JA. Patterns and processes in the California Current System. Progress in Oceanography. 2009;**83**:49-64. DOI: 10.1016/j.pocean.2009.07.028

[44] Mantua N, Hare S. The Pacific decadal oscillation. Journal of Oceanography. 2002;**58**:35-44. DOI: 10.1023/A:1015820616384

[45] Mann ME. On smoothing potentially non-stationary time series. Geophysical Research Letters. 2004;**31**:L07214, 4 pages. DOI: 10.1029/2004GL019569

[46] Curry JA, Webster PJ. Climate science and the uncertainty monster.

Bulletin of the American Meteorological Society. 2011;**3139**(1):1667-1682. DOI: 10.1175/2011BAMS3139.1

[47] Moore JC, Grinsted A, Jevrejeva S. New tools for analyzing time series relationships and trends. Eos, Transactions of the American Geophysical Union. 2005;**86**:226-232. Available from: https://www.glaciology.net/publication/2005-12-24-new-tools-for-analyzing-time-series-relationships-and-trends/

[48] Fraedrich K, Blender R. Scaling of atmospheric and ocean temperature correlations in observations and climate models. Physical Review Letters. 2003, 2003;**90**:108501-1-108501-4. DOI: 10.1103/PhysRevLett.90.108501

[49] Mantua NJ, Hare SR, Zhang Z, Wallace JM, Francis RC. A Pacific interdecadal climate oscillation with impacts on salmon production. Bulletin of the American Meteorological Society. 1997;**78**:1069-1079. Available from: http://citeseerx.ist.psu.edu/viewdoc/download?doi=10.1.1.553.1185&rep=rep1&type=pdf

[50] Breaker LC. A closer look at regime shifts based on coastal observations along the eastern boundary of the North Pacific. Continental Shelf Research. 2007;**27**:2250-2277. DOI: 10.1016/j.csr.2007.05.018

[51] Breaker LC, Flora SJ. Expressions of the 1976-77 and 1988-89 regime shifts in sea surface temperature off southern California and Hawai'i. Pacific Science. 2009;**63**:39-60. Available from: https://scripps.ucsd.edu/programs/shorestations/publications/

[52] Jo Y-H, Breaker LC, Tseng Y-H, Yeh S-W. A temporal multiscale analysis of the waters off the east coast of South Korea over the past four decades. Terrestrial, Atmospheric and Oceanic Sciences. 2014;**25**:415-434. DOI: 10.3319/TAO.2013.12.31.01

[53] Hickey BM. Then California Current System—Hypotheses and facts. Progress in Oceanography. 1979;**8**:191-279. DOI: 10.1016/0079-6611(79)90002-8

[54] Newman M, Gilbert GP, Alexander MA. ENSO-forced variability of the Pacific decadal oscillation. Journal of Climate. 2003;**16**:3853-3857. Available from: http://citeseerx.ist.psu.edu/viewdoc/download?doi=10.1.1.477.8989&rep=rep1&type=pdf

[55] Schneider N, Cornuelle BD. The forcing of the Pacific decadal oscillation. Journal of Climate. 2005;**18**:4355-4373. DOI: 10.1175/JCLI3527.1

[56] Shakun JD, Shaman J. Tropical origins of North and South Pacific decadal variability. Geophysical Research Letters. 2009;**36**:L19711, 5 pages. DOI: 10.1029/2009GL040313

[57] Enfield DB, Mestas-Nunez AM. Interannual to multidecadal climate variability and its relationship to global sea surface temperatures. In: Inter-Hemispheric Climate Linkages. Academic Press; 2001. pp. 17-29. ISBN: 9780124726703

[58] Gedalof Z, Smith DJ. Interdecadal climate variability and regime-scale shifts in Pacific North America. Geophysical Research Letters. 2001;**28**:1515-1518. Available from: https://www.uoguelph.ca/cedar/Pubs/gedalof_and_smith_GRL.pdf

Chapter 4

Significance of Mesoscale Warm Core Eddy on Marine and Coastal Environment of the Bay of Bengal

Nanda Kishore Reddy Busireddy, Kumar Ankur and Krishna Kishore Osuri

Abstract

Bay of Bengal (BoB) is an affluent region for the mesoscale (eddies) and synoptic scale (cyclones) systems. It occurs primarily through the seasonal variations, dynamical instabilities and equatorial wind forcing mechanisms. The individual or cumulative effect of these changes is vulnerable to the coastal and marine ecosystems. For example, tropical cyclone (TC) AILA experienced a warm core eddy (WCE) before the landfall, and consequently it intensified into a severe cyclonic storm (CS) and remained as a CS up to 15 h after the landfall. Its severity produces a heavy rainfall of >18 cm day^{-1}, thus leads to the coastal flooding. The eddy contribution to the TC is witnessed during and after the landfall. Inappropriately, high resolution *in-situ* observations are not available to identify such important processes on different time and spatial scales. Therefore, the present chapter analyses the northern BoB eddy induced signals using both *in-situ* and satellite (advanced microwave scanning radiometer—AMSR-2) derived products. Two *in-situ* locations (BD08 and BD09) are employed for this study purpose. The eddy responses at no-eddy, during and after eddy, have been analyzed. Besides, WCE imprints on the overlying atmosphere are also observed. The relationship between sea surface temperature and wind speed over the BoB region is assessed.

Keywords: warm core eddy, tropical cyclone, *in-situ*, AMSR-2, sea surface temperature, wind speed

1. Introduction

Marine environment usually influences or is influenced by the mesoscale (eddies) and synoptic scale (cyclones) processes. The horizontal scale of a typical mesoscale eddy is varying in between 100 and 200 km with a lifetime of 10–100 days [1–3]. The preserved shape carries/responsible for the mass or heat transport, turbulent current patterns, thermodynamic properties, potential fishing zones and upwelling processes over a long distance [3, 4]. Oceanic mesoscale eddies are of two types such as warm and cold core eddies. The significance of the warm (cold) core eddy is that it contains a higher (lower) temperature with elevated (lower) sea level in the center. The primary factors responsible for the genesis of the eddies were force exerted by the wind stress curl, meander separation, baroclinic instability and gradient wind balance [5]. Mesoscale eddies play an important role in governing the atmospheric vortices such as tropical cyclones (TCs) and ultimately affect its intensity. For example, cold core eddy promotes TC weakening while, warm core eddy (WCE) amplifies the TC intensity. It is known fact that the coastal and marine environments are sensitive to such oceanic vortices.

The Bay of Bengal (BoB) is an eddy rich region and is located in the northeastern part of the Indian Ocean (**Figure 1a**). The location of the BoB in the global and regional prospective is shown in the **Figure 1a**. It is dominated by the several factors such as freshwater discharge, seasonal occurrence of storms, monsoonal reversal winds and upwelling plumes [6, 7]. All these factors contribute to the formation of eddies throughout the year in this bay. The persistent eddy environmental characteristics determine the weather of the ocean and its overlying atmosphere. For example, ocean heat content is considered to be an important parameter in the ocean perspective that can modulate the atmospheric vortices [8, 9]. In case of WCE, the preserved heat content provides the sufficient positive energy to the storm in terms of large enthalpy fluxes and thus helps in the intensification process. Similarly, the process is opposite in case of cold core eddy. Moreover, it also has a

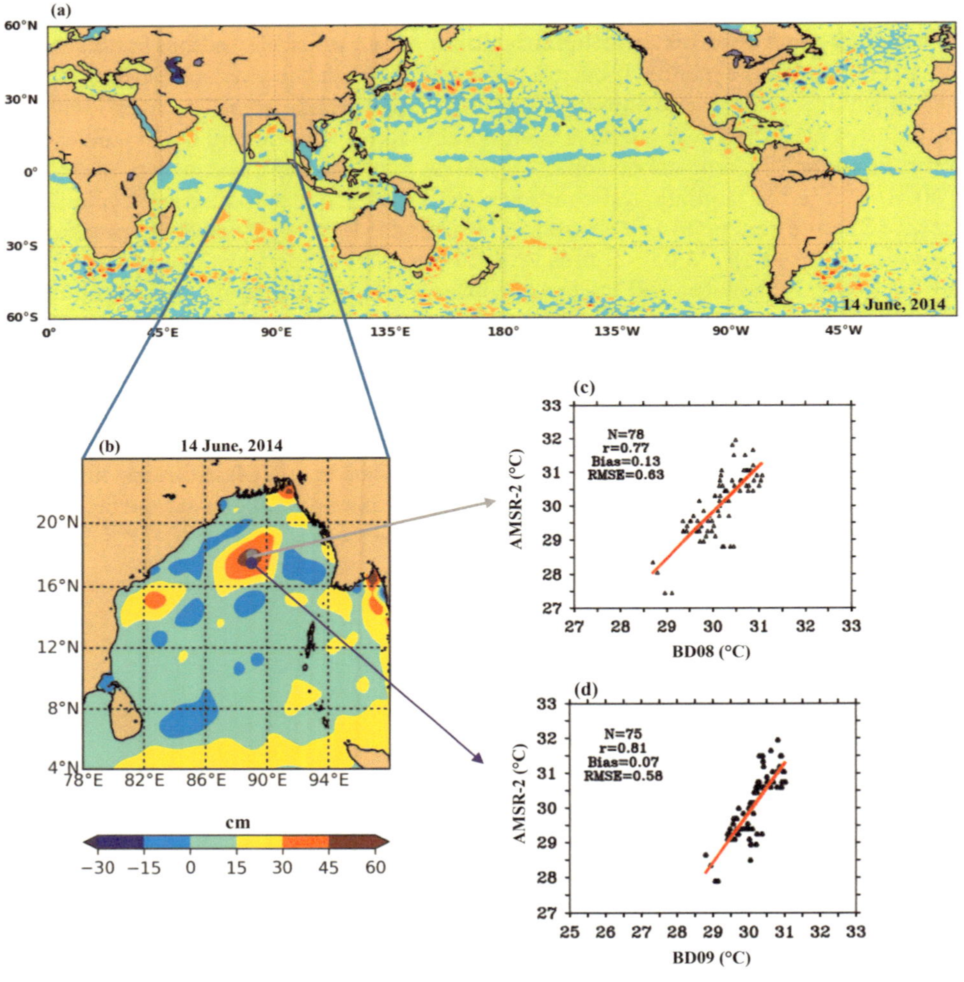

Figure 1.
Sea level anomaly of (a) global oceans, on 14 June 2014. The blue rectangle represents the zoomed version of (b) Bay of Bengal showing the warm core eddy. (c and d) The statistical analysis of sea surface temperature (°C) between AMSR-2 satellite and BD08 and BD09 observations during the eddy period (05 May–25 July 2014). Gray and blue color circles in (b) shows the BD08 and BD09 moorings.

major implication, when the WCE forms in the close proximity of the coastal region and a storm passes over on it. It poses a major socioeconomic impact on the coastal communities, agriculture and infrastructure, thus ultimately lead to the catastrophic damage. A best example related to this situation is TC Aila (2009) over the BoB. According to Regional Specialized Meteorological Center report (2010) and Sadhuram et al. [10], TC Aila suddenly gets intensified by 43% within a day before the landfall and caused for the 175 human deaths. These kind of similar events have been noticed around the global oceans. For example, Hurricane Opal (1995) in the North Atlantic Ocean and Super Typhoon Maemi (2003) in the western North Pacific under-went a WCE, and shows a sudden rapid intensification within a 24–36 h-time window [11, 12]. The principle behavior of these WCEs is that it acts like an insulating material against the ocean waters negative feedback and boost the TC intensity process [13]. While, it is true for cold core eddies thus provides a negative feedback to the storm and finally it gets weakens. However, there are other positive impacts associated with these cold core eddies that indirectly help the marine processes. It actually brings the subsurface chlorophyll maxima to the surface area and turn it into higher biological productivity in that region [14]. The detection of these eddies using the sea surface temperature (SST) gradient patterns is a difficult task because the attained surface variations are less. The better approach to deal with these mesoscale eddies are using the altimeter derived sea level anomalies (SLA) in which the high and the low SLA signal gives an indication of warm and cold core eddies [5].

Over the decade, understanding of these mesoscale eddy variability in the spatial and temporal manner has got a steady improvement. Bruce et al. [15] reported the mesoscale eddy features using the TOPEX altimeter data. Frenger et al. [16] analyzed the impact of these eddies on the atmospheric parameters such as winds, clouds and rainfall. Dandapat and Chakraborty [17] analysed the three dimensional structure of the mesoscale eddies using both the altimetry and Argo floats for the period 1993–2014. Busireddy et al. [18] shows the thermohaline structures associated with the WCE and its response on the overlying atmosphere. Despite these facts, the eddy-resolving models are still lacking to capture the realistic features of the ocean tractable in the BoB region. In summary, mesoscale eddies and its feedback to the atmospheric vortices suggest the urgency of high resolution ocean-atmosphere coupled models and satellite products over the region.

The present chapter is evolved by considering the availability of the existent high-resolution *in-situ* (buoy) and remote sensing (satellite derived) products and addresses the eddy induced signals in the ocean and atmospheric parameters. The overall discussion is followed primarily by choosing the (1) WCE during the normal conditions and in the presence of a cyclone environment, (2) sea surface temperature (SST) and wind speed data products verification of advanced microwave

Parameter	Sensor type	Resolution	Accuracy	Range
In-situ SST	Thermistor/Seabird Micro-CAT SBE37	0.0001°C	0.002°C	−5 to 35°C
In-situ wind speed	Vane + flux gate compass/ Lambrecht	0.1 ms^{-1}	±2%	0 to 35 ms^{-1}
AMSR-2 SST	Microwave radiometer	0.25° × 0.25°	±0.8°C (released)	−2 to 35°C
AMSR-2 wind speed			±1.5 ms^{-1} (released)	0 to 30 ms^{-1}

Table 1.
Specifications of SST (°C) and wind speed (ms^{-1}) parameters of both in-situ and AMSR-2 sensors.

scanning radiometer (AMSR-2) over the BoB region and (3) SST-wind speed relationships during the WCE, near coastal region and open oceans using satellite and *in-situ* measurements. This study signifies the inevitability of high resolution products in the coastal and marine environments over the BoB. The resolution and accuracy of the satellite and *in-situ* products is demonstrated in **Table 1**. The buoys (BD08 and BD09) collects both the surface met and ocean parameters with depth. The thermohaline features are nearly similar at both the locations before and after the warm core eddy period. However, significant differences are seen during warm core eddy period owing to its spatial variation. The vertical thermohaline structures exhibited ridge and trough structures throughout the eddy life. It is worthwhile to note that the structures are prominent during developing stage of warm core eddy, however, they are nearly uniform in the weakening phase of warm eddy. More details on the eddy induced signals before, during and after the warm core eddy can be obtained from Busireddy et al. [18].

2. North Bay of Bengal

The Bay of Bengal is surrounded by India (in the west and northwest side), Bangladesh (in the north), Myanmar and the Andaman/Nicobar Islands of India (in the east) and Sri Lanka (in the south) (**Figure 1a**). A number of major rivers belonging to the India, Bangladesh and Myanmar is flowing into the northern BoB. In the tropics, Northern BoB has a unique feature among the world as it is dominated by the low surface salinity due to seasonal reversal of the monsoon winds, freshwater discharge from the major rivers and the seasonal occurrence of cyclones [19]. During the summer monsoon season (June-September), northern BoB is conducive for the several meteorological systems that brings rainfall to the central and Northern parts of India. Alike, the massive river outflow, which is coupled with rainfall eventually drain into the bay region. These large supplies of fresh water and rainfall resulting in the strong near surface layer stratification led to the formation of the barrier layer. Likewise, during winter season (November-December), the region experiences a strong surface layer temperature inversion due to the sustained salinity stratification of the summer monsoon. Moreover, the region is active for the mesoscale eddies that are formed due to the active coastal upwelling Kelvin wave during the spring intermonsoon period (March-May) [18, 20].

2.1 Warm core eddy

A warm core eddy has been identified using the spatial maps of sea level anomalies (SLA) obtained from the Archiving, Validation, and Interpretation of Satellite Oceanographic (AVISO) over the northern BoB (**Figure 1b**). It is noted that the size of the eddy is around 5° × 5°. Chelton et al. [21] criteria are applied for the eddy identification purpose and realized that it is centered at around 89°E and 18°N with a life period of 3 months (05 May 2014–25 July 2014) in the northern BoB. The eddy was surrounded by the several alternative warm and cold core rings (**Figure 1b**). The further details regarding the genesis and its propagation characteristics are demonstrated in Busireddy et al. [18]. In general, the eddy size is often associated with the Rossby radius of deformation which is increased from higher to lower latitudes. It is termed as the length scale at which rotational effects become equally important as the buoyancy or gravitational effects in a flow field. From Figure 3b of Chen et al. [22], the Rossby radius of deformation values are 60 km at 20°N and 130 km at 9°N. Incidentally, two of the moored buoys (BD08 and BD09), deployed earlier by National Institute of Ocean Technology, are located within the eddy region. The buoys provide the incessant time series measurements of ocean temperature and salinity profiles. Therefore, the estimated Rossby radius (L = NH/f) at 18°N latitude using the buoy observations is 39 km and is consistent with Chen et al. [22]. Here, N is the Brunt-Vaisala frequency (0.0179 s^{-1}), H is the scale height or equivalent depth (100 m), and f is Coriolis frequency (4.505×10^{-5} s^{-1}). According to the Space Application Centre, Indian Space Research Organization (ISRO), India (https://www.mosdac.gov.in/oceanic-eddies-detection), the radius of the present eddy derived from AVISO data is ~170 km. The fact to be noted that the eddy radius (~170 km) is much bigger than the Rossby radius (39 km) means that there are some other effects contributing to this large-scale eddy formation.

The statistical analysis of SST between the *in-situ* and AMSR-2 is carried out at two moored buoy locations in the northern BoB (BD08 at 89°E/18°N and BD09 at 89°E/17.5°N) to quantify the uncertainty of the satellite derived SST in the presence of WCE (**Figure 1c** and **d**). The AMSR-2 data description and its validation information are made available online at http://www.remss.com/missions/amsr/. Considering the surface ocean temperatures, the buoys observed an unusual increase of 3–4°C during the WCE period. In order to capture these kinds of similar variations using the satellite, primarily, the capability of the AMSR-2 SST is verified with respect to the BD08 and BD09 during the eddy period. The statistical analyses show a root mean square error (RMSE) of 0.5–0.6°C at both the buoy locations. The correlation and bias are varying in between 0.7–0.8 and 0.07–0.13°C, respectively (**Figure 1c** and **d**). As stated in Reddy et al. [23], the AMSR-2 mean SST error over the BoB is limited to ~0.3°C. In this situation, the observed RMSE (~0.5–0.6°C) in the WCE period is doubled as that of mean RMSE (~0.3°C) over the BoB. It infers that AMSR-2 SST shows a large error in the presence of mesoscale eddy. However, these error values are within the satellite prescribed release accuracy range (i.e., ±0.8°C). The results provide that AMSR-2 is over estimating for the higher SST values.

In addition to the ocean responses, the WCE provides it signatures on the surface meteorological variables. The noticeable variations on the surface-air temperature (SAT) and sea-surface temperature (SST) anomalies, air-sea specific humidity values and enthalpy fluxes during the WCE period is observed [18]. During the eddy progression, i.e., from genesis to peak stage, both the SAT and SST anomalies show similar changes and exhibits its peak magnitude of 3°C at the peak stage. In contrast, significant differences have been noticed during the peak-to-dissipation stage. The mean differences between the SAT and SST anomalies are 0.01 and 0.6°C, respectively. These differences indirectly cause for the changes in the surface turbulent fluxes, which is the main source for weather systems (i.e., cyclones). It is observed that the strong air-sea temperature differences in the WCE lead to the sharp increase in the latent heat flux and sensible heat flux of 504 and 142 W m^{-2}, respectively [18]. The comparison of these values with the climatology revealed that the eddy induced fluxes are higher by 274% and 370% [7, 24]. **Table 2** illustrates the estimated values of buoy (BD08) air-sea parameters and its corresponding AVISO SLA values during the eddy evolution. In order to understand clearly, the WCE period is broadly classified into various phases such as no-eddy, genesis, intensifying phase, peak stage, phase and dissipation phase, respectively [18]. The obtained results indicated that for all the parameters, values are increasing beginning from the no-eddy, genesis and reaches to its higher values at the peak stage. Thereafter, it reaches to its normal values during the dissipation stage. For example, the estimated ocean heat content at peak stage shows an increment of 258%, when compared to the no-eddy conditions [18]. Similarly, the other parameters such as wind speed, SST and SLA also exhibits a similar increment of 51, 3.6 and 98%, respectively when compared to that of no-eddy conditions.

2.2 Role of warm core eddy in the intensification of AILA (May 2009) cyclone over BoB

The greatest contributors to the coastal flooding and its associated damage are mainly due to the landfalling cyclones. It is generally formed over the warm waters (>26.5°C) and gains the remarkable amount of energy from the ocean [25]. The typical size of the TCs are 200–2000 km with a time period of 1–2 weeks [26]. Aila is a severe cyclonic storm that formed over the northern BoB and made landfall near the Sagar Island on 25 May 2009. The strong winds and storm surge cause for the 175 human deaths (http://www.rsmcnewdelhi.imd.gov.in). The prominence of cyclone is that it intensified from cyclone to severe cyclonic storm within a few hours before the landfall. The later studies revealed that one of the reasons for sudden intensification is due to the presence of WCE in near at the coastal region of the West Bengal. **Figure 2** shows the AILA cyclone track and its corresponding time series analysis of SLA, cyclone intensity and wind intensity (kt) values. The SLA depicts the occurrence of WCE close proximity to the West Bengal coast. During the Aila cyclone movement, it encountered a WCE on 24 May 2009 as a cyclonic storm and get intensified into a severe cyclonic storm within a short time span. It is observed that WCE maintains an SST of 31°C during the time and favors the storm intensity increase by 43%. The high ocean heat content and deep isothermal warm

Phases/parameters	OHC (kJ cm^{-2})	Wind speed (ms^{-1})	SST (°C)	SLA (cm)
Before/no-eddy	40	4.9	30.0	25.4
Genesis	69	1.4	30.1	26.1
Intensifying phase	109	4.7	30.4	38.3
Peak stage	139	7.4	31.1	50.5
Weakening phase	100	7.2	29.7	35.8
Dissipation stage	7	5.0	28.6	13.7

Table 2.
Quantification measures of BD08 atmosphere-ocean parameters during no-eddy to dissipation stages.

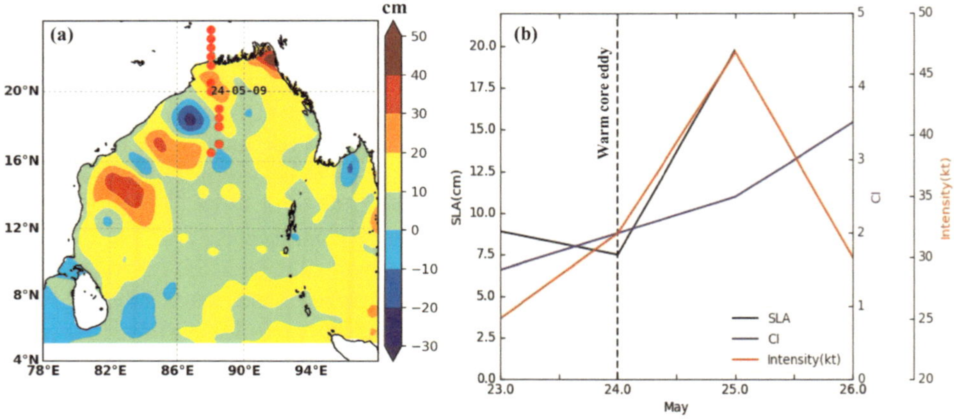

Figure 2.
AILA tropical cyclone passing through the warm core eddy in (a) and its corresponding time series analysis of sea level anomaly (SLA), cyclone intensity (CI) and wind intensity (kt) values along the track shown in (b). Shaded background in (a) and vertical dashed line in (b) represents the SLA and warm core eddy crossing day respectively.

DOI: http://dx.doi.org/10.5772/intechopen.86243

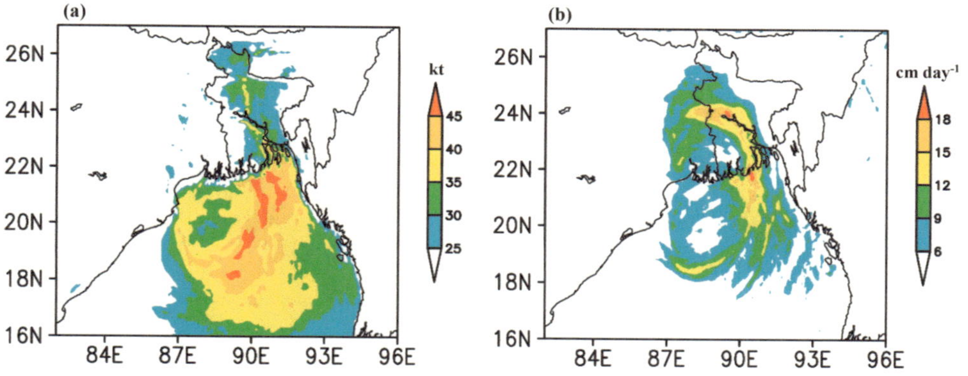

Figure 3.
(a) Maximum sustained wind swath in the life span of the TC AILA (23–26 May 2009) and (b) 24-h accumulated rainfall during landfall day valid at 03 UTC 26 May 2009.

layer within the eddy inhibit the storm induced cooling and further help in the intensification process [10]. From **Figure 2b**, the time series analysis of SLA and wind intensity shows a usual value ranging from 7.5 to 9 cm and 25–32 knots before it arrives into the eddy region. Once the TC encountered the eddy (24 May 2009), SLA shows a sudden increment of 9–20 cm with the corresponding increase in wind intensity from 32 to 47 knots. Sadhuram et al. [10] explained that high SST and large enthalpy fluxes beneath the eddy provides a positive feedback and support to the TC intensification process. Along with this, cyclone intensity also shows a positive trend throughout the TC life period. Moreover, the cyclone made landfall on 25 May 2018 so the SLA is absent and wind intensity values decreases due to the land interaction.

Figure 3 shows the spatial distribution of wind speed and rainfall associated with the Aila cyclone during the landfall. The wind speed was obtained from FiNaL (FNL) analyses of the National Center for Environmental Prediction (NCEP) and the rainfall (3-hourly) from the Tropical Rainfall Measuring Mission. The structural wind swath analyses show a peak wind intensity of >45 knots that persists over the Bangladesh even after the landfall. As mentioned earlier, the presence of WCE during the AILA passage supports its intensification into a severe cyclone, while the abundant soil moisture from the land surface conditions (Delta region of the River Ganga) helps to sustain/maintain the intensity for next 15 h. The analysis from the spatial distribution of 24-h accumulated rainfall during the landfall exhibits a high rainfall of ~15–18 cm day^{-1}. Moreover, these rainfall are elongated to a few hundred kilometers away from the coast. The down-pouring of heavy rain in some places around the Sundarban area lead to the coastal inundation and flooding. According to the India Meteorological Department, New Delhi, a storm surge of ~3 m is experienced in the western regions of Bangladesh and it is 2–3 m over the Sundarban area region (http://www.rsmcnewdelhi.imd.gov.in).

3. SST and wind speed analysis over the BoB

As we know that SST is a vital parameter in the coastal and marine environment that can regulate both the physical and biological processes. The SST changes are directly associated with the magnitude of the wind speed. Higher (lower) the magnitude lower (higher) the SST values. Here, the performance of the AMSR-2 SST and wind speed is verified at two *in-situ* locations (BD12—94°E, 10.3°N and BD14—88°E, 7°N) over the BoB for the period 2013–2014. Out of these two buoys,

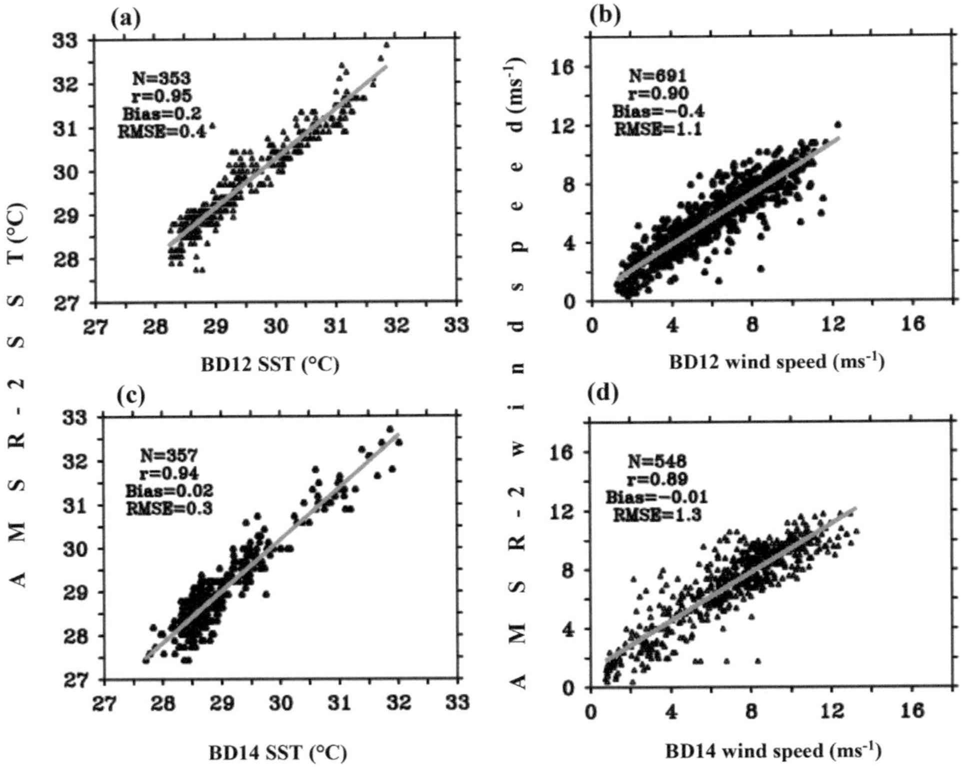

Figure 4.
Scatter plot analysis of AMSR-2 SST at (a) BD12—coastal region and (c) BD14—open ocean. (b) and (d) are same as (a) and (c) but for the wind speed during the period 2013–2014.

BD12 is situated near at the coastal region, i.e., east of the Andaman Island (~100–150 km) and the BD14 is placed in the open ocean, i.e., southern BoB region. **Figure 4** shows the statistical analyses of AMSR-2 SST and wind speed with respect to the *in-situ* observations. The formulations used in the present study are described in the Section 3.1. The SST statistics revealed that the BD12 and BD14 show an RMSE (Eq. (1)) error of 0.4 and 0.3°C (**Figure 4a** and **c**). Considering its correlation (r) and bias, BD12 (BD14) shows 0.95 (0.94) and 0.2°C (0.02°C), respectively (Eqs. (2) and (3)). It means that expected AMSR-2 accuracy is less towards the coastal region as compared to the open ocean. The similar statistics are conducted for the satellite derived wind speed. The results showed that an RMSE of 1.1 and 1.3 m s^{-1}, respectively, with a bias and correlation of 0.9 m s^{-1} and −0.4 (−0.02) at BD12 and BD14 locations (**Figure 5b** and **d**). However, all these obtained SST and wind speed errors are within the prescribed accuracy mentioned by the AMSR-2 (±0.8°C and ±1.5 m s^{-1}).

3.1 Statistical formulation used

A common set of mathematical and statistical techniques has been used to assess a performance of the AMSR-2 satellite measurements over the BoB region. It could further helpful to quantify the uncertainty and as well as for the rational use of the AMSR-2 information over the current region. The study employs the most widely used metrics such as bias, linear correlation and root mean square error, respectively. The formulas related to the error statistics are as follows.

$$RMSE = \sqrt{\frac{1}{n}\ \sum_{i=1}^{n}((x_i - x_{obs})/x_{obs})^2} \quad (1)$$

$$Bias = x_i - x_{obs} \quad (2)$$

$$correlation\ (r) = \frac{1}{n-1}\sum(x_i - \overline{x}/s_x)\ (y_i - \overline{y})/s_y) \quad (3)$$

where n, x_i, and x_{obs} represents the total number of samples, satellite and *in-situ* observations, respectively. $\overline{x}$ and $\overline{y}$ indicates the temporal mean of satellite and *in-situ* measurements and whereas s_x and s_y shows the standard deviation of it.

3.2 SST—wind speed relationship over the BoB

This section briefly explains the SST dependency on the surface wind speed over the BoB region (**Figure 5**). The analysis comprises of northern to southern BoB at different locations and scenarios such as eddy environment (BD08), near a coastal region (BD12) and open ocean areas (BD10 and BD14). Note that the SST-wind relationship is evaluated at BD08 location only during the WCE period. Because, the

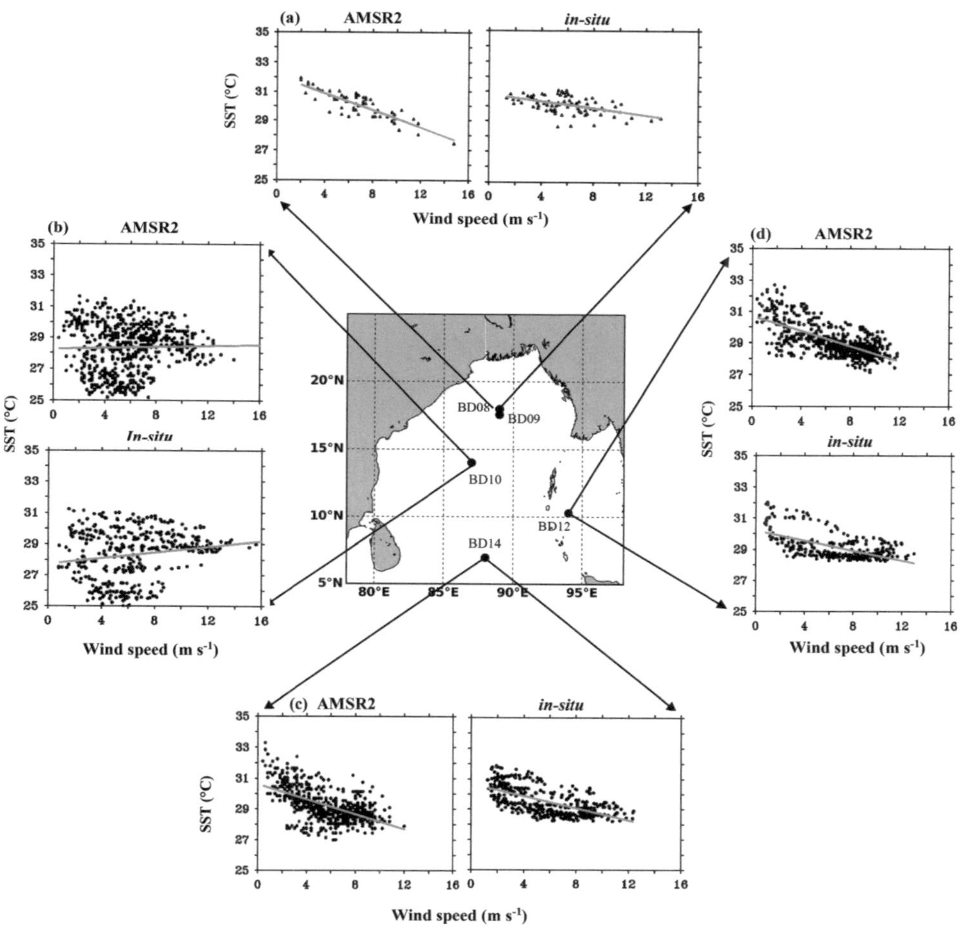

Figure 5.
Relationship between SST (°C) and wind speed (m s^{-1}) of AMSR-2 and in-situ *measurements during (a) warm core eddy period (b and c) at BD10 and BD14—open ocean and (d) BD12—coastal region. Note that warm core eddy period considers only the BD08 observations.*

estimated variations are minimal at both the BD08 and BD09 so the analysis related to the BD09 is excluded here. **Figure 5a** shows the SST-wind speed relation during the WCE period. The relationship shows a negative trend inferring that SST decreases with an increase in wind speed for both AMSR-2 and *in-situ* observations. The observed slope of the fitted line is high for the AMSR-2 with an SST difference of 3.4°C and is limited to 2°C in case of BD08. However, the long time period (2013–2014) analyses at the same location shows a contradictory relation indicating SST increases with an increase in wind speed [23]. **Figure 5b** shows the SST-wind speed relation over the central BoB region, i.e., BD10. Here, the relation displays a positive linear trend, addressing SST increases with an increase in wind speed [23]. Moreover, the deviation of SST and wind speed values of the fitted line is higher (lower) for the low (high) wind conditions. The data spread could be due to the fact that this region is active for the seasonal storms and monsoons. Considering the same relationship over the BD12 (eastern BoB) and BD14 (southern BoB), the SST decreases with an increase in wind speed (**Figure 5c** and **d**). It means that higher (lower) SST values are corresponding to the lower (higher) wind speed values. The present analysis also revealed that obtained AMSR-2 fitted line slope is relatively larger than the buoys fitted line. It could be due to the positive bias of SST and negative bias of wind speed over the BoB. In summary, satellite is underestimating the SST and wind speed variations because the ocean characteristics changes with the season and latitude over the BoB.

4. Conclusions

This chapter addresses the mesoscale warm core eddies and high-resolution products prominence in the Bay of Bengal. *In-situ* observations are used to analyze and quantify the errors of the AMSR-2 during the WCE over the northern BoB. The eddy has a life period of ~81 days with a 5° × 5° size. The statistical analysis of AMSR-2 SST with *in-situ* observations during the eddy period show an RMSE of 0.5–0.6°C. Interestingly, the obtained SST error is double that of the mean error calculated over the BoB region. In addition, the eddy impact on the overlying atmosphere is analyzed with showing an increase in the enthalpy fluxes and surface atmospheric parameter values. The similar event has been undertaken in the presence of AILA cyclone and the eddy role in the intensification process is observed. Briefly, Aila encounters the WCE on 24 May 2009 and after spending a few hours, it gets intensified into a severe cyclonic storm. The time series analysis of satellite and met-ocean parameters depicted that SLA and wind intensity shows its peak value of 20 cm and >45 knots after it passes through the WCE. Due to this sudden intensification, the coastal region receives a heavy rainfall (>18 mm day^{-1}) and strong winds (>45 knots) that led to the coastal inundation and flooding problems.

The AMSR-2 SST and wind speed parameters are evaluated qualitatively and quantitatively with respect to the buoy observations over the BoB. The statistical analysis of SST at BD12 and BD14 locations show an RMSE of 0.4 and 0.3°C and whereas wind speed shows an RMSE of 1.1 and 1.3 m s^{-1}, respectively. Along with this, the AMSR-2 SST and wind speed relationship is assessed in different aspects such as WCE, open and near coastal regions of the BoB. These analyses cover the whole BoB region and the obtained SST and wind speed interactions helpful to understand the air-sea interaction processes. The relationship exhibits a both linear positive and negative slopes over the BoB. It infers that wind speed and SST shows contrasting features within the BoB region. For example, BD08, BD12 and BD14 displays the negative relation inferring SST decreases with an increase in wind speed in both AMSR-2 and buoy observations. However, BD10 location shows the opposite relation (i.e., positive relation) indicating SST increases with an increase in the wind speed. The overall results directed that AMSR-2 is able to replicate the buoy induced signals with positive bias in the SST and negative bias in the wind speed over the BoB.

Acknowledgements

The authors gratefully acknowledge the financial support of SERB (ECR/2016/001637), and the SERB—Purdue Overseas Visiting Doctoral Fellowship (OVDF) scheme (SB/S9/Z-03/2017), Govt. of India. The authors owe thanks to the Department of Space (DOS), Govt. of India, for providing the warm core eddy features.

Conflict of interest

The authors declare that they do not have any competing interests to publish this book chapter.

List of acronyms

AMSR-2	advanced microwave scanning radiometer-2
AVISO	archiving, validation, and interpretation of satellite oceanographic
BoB	Bay of Bengal
RMSE	root mean square error
SAT	surface air temperature
SLA	sea level anomalies
SST	sea surface temperature
TC	tropical cyclone
WCE	warm core eddy

Author details

Nanda Kishore Reddy Busireddy[1,2], Kumar Ankur[1] and Krishna Kishore Osuri[1]*

1 Department of Earth and Atmospheric Sciences, National Institute of Technology Rourkela, Odisha, India

2 Department of Earth, Atmospheric, and Planetary Sciences, Purdue University, West Lafayette, IN, USA

*Address all correspondence to: osurikishore@gmail.com

References

[1] Patnaik KVKRK, Maneesha K, Sadhuram Y, Prasad KVSR, Ramana Murty TV, Brahmananda Rao V. East India Coastal Current induced eddies and their interaction with tropical storms over Bay of Bengal. Journal of Operational Oceanography. 2014;**7**(1): 58-68

[2] Murty TV, Rao MM, Sadhuram Y, Sridevi B, Maneesha K, Sujith Kumar S, et al. Objective mapping of observed sub-surface mesoscale cold core eddy in the Bay of Bengal by stochastic inverse technique with tomographically simulated travel times. Indian Journal of Geo-Marine Sciences. 2011;**40**:307-324

[3] Faghmous JH, Frenger I, Yao Y, Warmka R, Lindell A, Kumar V. A daily global mesoscale ocean eddy dataset from satellite altimetry. Scientific Data. 2015;**2**:150028

[4] Beron-Vera FJ, Wang Y, Olascoaga MJ, Goni GJ, Haller G. Objective detection of oceanic eddies and the Agulhas leakage. Journal of Physical Oceanography. 2013;**43**(7):1426-1438

[5] Gopalan AKS, Krishna VV, Ali MM, Sharma R. Detection of Bay of Bengal eddies from TOPEX and in situ observations. Journal of Marine Research. 2000;**58**(5):721-734

[6] Chen G, Wang D, Hou Y. The features and interannual variability mechanism of mesoscale eddies in the Bay of Bengal. Continental Shelf Research. 2012;**47**:178-185

[7] Kumar BP, Vialard J, Lengaigne M, Murty VSN, Mcphaden MJ. TropFlux: Air-sea fluxes for the global tropical oceans—Description and evaluation. Climate Dynamics. 2012;**38**(7-8): 1521-1543

[8] Goni GJ, Trinanes JA. Ocean thermal structure monitoring could aid in the intensity forecast of tropical cyclones. Eos, Transactions American Geophysical Union. 2003;**84**(51): 573-578

[9] Mainelli M, DeMaria M, Shay LK, Goni G. Application of oceanic heat content estimation to operational forecasting of recent Atlantic category 5 hurricanes. Weather and Forecasting. 2008;**23**(1):3-16

[10] Sadhuram Y, Maneesha V, Murty TVR. Intensification of Aila (May 2009) due to a warm core eddy in the north Bay of Bengal. Natural Hazards. 2012; **63**(3):1515-1525

[11] Oropeza F, Raga GB. Rapid deepening of tropical cyclones in the northeastern Tropical Pacific: The relationship with oceanic eddies. Atmosfera. 2015;**28**(1):27-42

[12] Hong X, Chang SW, Raman S, Shay LK, Hodur R. The interaction between Hurricane Opal (1995) and a warm core ring in the Gulf of Mexico. Monthly Weather Review. 2000;**128**(5): 1347-1365

[13] Wu CC, Lee CY, Lin II. The effect of the ocean eddy on tropical cyclone intensity. Journal of the Atmospheric Sciences. 2007;**64**(10):3562-3578

[14] Prasanna Kumar S, Muraleedharan PM, Prasad TG, Gauns M, Ramaiah N, De Souza SN, et al. Why is the Bay of Bengal less productive during summer monsoon compared to the Arabian sea. Geophysical Research Letters. 2002; **29**(24):88-81

[15] Bruce JG, Kindle JC, Kantha LH, Kerling JL, Bailey JF. Recent observations and modeling in the Arabian Sea Laccadive high region. Journal of Geophysical Research, Oceans. 1998;**103**(C4):7593-7600

[16] Frenger I, Gruber N, Knutti R, Münnich M. Imprint of Southern Ocean eddies on winds, clouds and rainfall. Nature Geoscience. 2013;**6**(8):608

[17] Dandapat S, Chakraborty A. Mesoscale eddies in the Western Bay of Bengal as observed from satellite altimetry in 1993–2014: Statistical characteristics, variability and three-dimensional properties. IEEE Journal of Selected Topics in Applied Earth Observations and Remote Sensing. 2016;**9**(11):5044-5054

[18] Busireddy NKR, Osuri KK, Sivareddy S, Venkatesan R. An observational analysis of the evolution of a mesoscale anti-cyclonic eddy over the Northern Bay of Bengal during May–July 2014. Ocean Dynamics. 2018; **68**(11):1431-1441

[19] Thadathil P, Muraleedharan PM, Rao RR, Somayajulu YK, Reddy GV, Revichandran C. Observed seasonal variability of barrier layer in the Bay of Bengal. Journal of Geophysical Research-Oceans. 2007;**112**(C2):1

[20] Nuncio M, Prasanna Kumar S. Evolution of cyclonic eddies and biogenic fluxes in the northern Bay of Bengal. Biogeosciences Discussions. 2013;**10**(10):16213-16236

[21] Chelton DB, Schlax MG, Samelson RM, de Szoeke RA. Global observations of large oceanic eddies. Geophysical Research Letters. 2007;**34**(15):p. L15606. DOI: 10.1029/2007GL030812

[22] Chen G, Hou Y, Chu X. Mesoscale eddies in the South China Sea: Mean properties, spatiotemporal variability, and impact on thermohaline structure. Journal of Geophysical Research, Oceans. 2011;**116**(C6):p.C06018

[23] Reddy BNK, Venkatesan R, Osuri KK, Mathew S, Kadiyam J, Joseph KJ. Comparison of AMSR-2 wind speed and sea surface temperature with moored buoy observations over the Northern Indian Ocean. Journal of Earth System Science. 2018;**127**(1):14

[24] Cyriac A, Ghoshal T, Shaileshbhai PR, Chakraborty A. Variability of sensible heat flux over the Bay of Bengal and its connection to Indian Ocean Dipole events. Ocean Science Journal. 2016;**51**(1):97-107

[25] Gray WM. Global view of the origins of tropical disturbances and storms. Monthly Weather Review. 1968; **96**:669-700

[26] Niyogi D, Subramanian S, Osuri KK. The role of land surface processes on tropical cyclones: Introduction to Land Surface Models. In: Advanced Numerical Modeling and Data Assimilation Techniques for Tropical Cyclone Prediction. Dordrecht. Springer; 2016. pp. 221-246

Section 3

Coastal Processes and HF Communications in Coastal and Marine Environments

Chapter 5

Modeling of Coastal Processes in the Mediterranean Sea: A Pilot Study on the Entrance of Suez Canal in Egypt

Mona Fouad Kaiser, Walaa Awaad Ali and Maysara Khairy El Tahan

Abstract

The main objective of this research is applying numerical modeling to simulate the impact of the Suez Canal jetties on the beach morphology and hydrodynamic regime along the Suez Canal coastal zone. In addition, coastal processes including waves and wave-induced currents will be evaluated using 2D modeling. This research will contribute to quantify the shoreline stability during the last three decades. Hydrodynamic and sediment transport (ST) models are utilized to predict sediment transport pathways and how sediment might move within the entrance of Suez Canal port. Remote sensing analyses of the Landsat Thematic Mapper images during 2000–2018 show siltation processes at the entrance of the Suez Canal. Vector analyses of the images' data indicated updrift accretion at a rate of +15 m/year and downdrift erosion at a rate of −13 m/year. Coastal processes including waves and currents contribute to shoaling problem along the navigation channel of the Suez Canal port. Applications of 2-3D models were used to simulate wave and current dissipation. In addition, beach slope profiles and hydrodynamic models are used to help in understanding the impact of coastal processes on beach morphology and hydrodynamic regime controlling siltation problem along the entrance of Port Said harbor.

Keywords: 2D modeling, Mediterranean Sea, coastal processes, Suez Canal port, hydrodynamic regime, beach morphology

1. Introduction

The Suez Canal is located in Egypt west of the Sinai Peninsula. Its construction was preceded by cutting a small freshwater canal from the Nile Delta and connecting it with a southern branch to Suez and a northern branch to Port Said. The Suez Canal is considered to be the first artificial canal to be used in Travel and Trade. It is completed to create the first saltwater passage between Port Said on the Mediterranean and Port of Suez on the Red Sea, providing an essentially direct route for transport of goods and petroleum tankers between Europe and Asia. The construction of Suez Canal, nearly from 40 centuries, by the pharaohs, aims at to do linking between the Red Sea and Mediterranean Sea. In addition, Suez Canal Authority is responsible to do periodical dredging for the navigation channel and its surroundings to keep this channel deep and safe. The canal supports approximately 8% of the world's shipping traffic with almost 50 vessels traveling through the canal daily. It has 195 km length; its width ranges from 60 to 300 m. It is able to accommodate ships as large as 150,000 tons fully loaded (Suez Canal Authority personal communication). This study aims to understand the main factors controlling siltation problem in the entrance of the Suez Canal port. Numerical modeling will be used to simulate coastal processes, beach profiles, and hydrodynamic regime. The results help in shoaling mitigation and facilitating passing of high loading ships along the canal.

As the important geographical location of Port Said Governorate, it has many activities in national and regional development. In addition, it is considered as the Gulf of Suez extension. Consequently, it has valuable resources such as the Mediterranean Sea, the Red Sea beaches, lakes, protected areas, and historical and archeological areas. These resources are suitable for tourism development. Therefore, Port Said has quickly become the third largest urban governorate in Egypt with respect to population.

2. Study site description

The Suez Canal coastal zone lies between longitudes 32°13′ and 32°25′ E and between latitudes 31°10′ and 31°20′ N (**Figure 1**). The concerned site represents a part of the Egyptian Mediterranean coast lying to the north of the Nile Delta east of Port Said. The beach profile slope has 1 m/km, and the depth of seabed reaches 25 m at the northern boundary of the study site. The beach sediment, along the coastal zone, is mainly composed of sand; its limit reaches 5 m in depth [1]. Although going to the sea bed offshore, the sediment texture that is covering the seabed was changed from muddy sand in the area limit between 5 and 10 m to muddy in deeper zone. Abu Asi [2] concluded that the coastal zone of Sinai from Port Said to El Arish is under extensive development. Consequently, several integrated development projects are being implemented along the coastal zone of North Sinai including the El-Tina plain [3]. The area is identified as it is completely covered by quaternary sediments of littoral, alluvial, and eolian origin, which show variations in their texture and composition ranging from unconsolidated sands to salinized silt and clay of chemical and biochemical origins. They also described the area as it has a concave shoreline configuration that is about 39 km long and 818 km^2 in area. The plain is subsiding at a rate of about

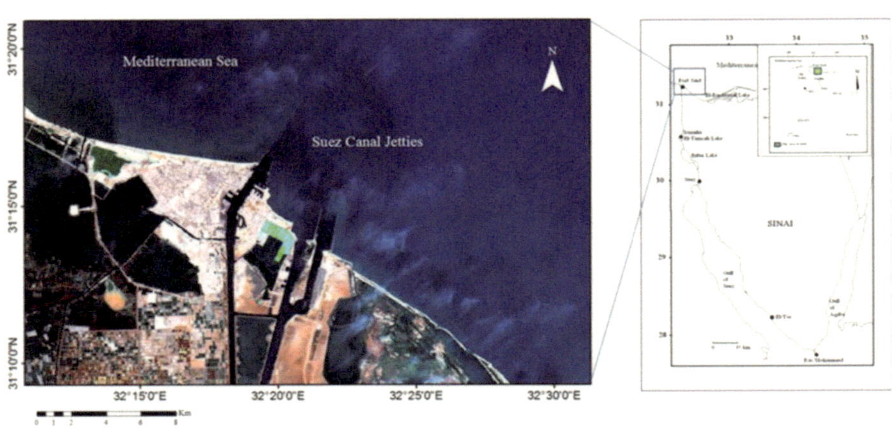

Figure 1.
Suez Canal entrance; protected by eastern and western jetties.

0.5 cm/year. The only engineering structures built at the study area are the 7.7 and 2.0 km jetties constructed to protect the inlet at Port Said and the East Port Said harbors, respectively. Additionally, the thickness of Holocene strata beneath the modern delta plain is a direct function of subsidence, which ranges from 50 m at Port Said and tends to decrease or be nearly absent westward below the Alexandria coastal plain.

The principal transporting agents in the concerned site are waves and wave-induced longshore current [4]. The wave rose was constructed based on records measured between 1997 and 1999 off the Damietta Harbor using a pressure wave gauge (InterOcean System S4DW) installed at ~12 m water depth [5]. The average significant wave height ranges from 1.04 to 4.45 m with long duration, its direction is mainly coming from NW in winter. These waves are responsible for generating the longshore currents and transporting sediment toward the east. However, the N-E waves having short duration are responsible for generating a reverse longshore current toward the west [6].

The eroded sediment of Damietta promontory was blocked west of Port Said causing accretion of +15 m/year along the western jetty of the El-Gamil inlet. Growth of tombolos occurred behind detached breakwaters at a rate of +6 m/year. The resulting break in longshore drift caused erosion of −6 m/year downdrift of the breakwaters. The eastern side of the Suez Canal, Bur Fouad, is suffering from erosion at a rate of −18 m/year. The coastline of El-Bardawil Lake is experiencing accretion of +6 m/year in some sections and erosion of −9 m/year in others [7]. The basins inside the Port Said harbor have a depth ranging from 8 to 24 m water; it is subjected to a severe sand drift. Suez Canal Authority usually keeps it clear by dredging. Maintenance dredging is simply the removal of sediments from a body of water that have accumulated due to erosion in order to maintain a desired depth, as in a navigation channel. Suez Canal navigation channel is authorized to be maintained to certain depths depending on its use, by periodic dredging of the silt, sand, and clay that are deposited in it [8].

In order to evaluate the impact of engineering protection on the coastal processes including waves and currents and beach profile configuration, numerical modeling techniques are utilized to predict the patterns of shoreline changes due to the changes in wave conditions.

3. Remote sensing techniques and results

3.1 Change detection

Shoreline positions were obtained from the TM band-7 images using a region-based segmentation process in which the sea area was extracted as a region [9]. Region growing techniques are generally better in noisy images, where borders are extremely difficult to delineate. Homogeneity is an important character of regions and is used as the main segmentation criterion in region growing. Thematic Mapper band 7 (short-wave infrared) was used for the image segmentation procedure to produce a vector map of the shoreline, so that the land-sea boundary could be delineated. Shorter wavelengths can pass through shallow water, making accurate delineation of the coastline difficult [10]. Using short-wave infrared data ameliorates the high-reflectance problems caused by surf in the breaker zone [11]. A line representing shoreline position (the boundary between sea and land) was created along the Suez Canal coastal zone. The output data were saved as a vector file enabling analysis of coastline change using geographic information system (GIS) software [12]. Shoreline displacements during the 2000–2018 period were extracted from the images using the measurement tools in ERDAS IMAGINE VirtualGIS. Edge detection and segmentation seem to be the most suitable approach to produce vector map data for the study site. The results indicate updrift accretion at a rate of +15 m/year and downdrift erosion at a rate of −13 m/year along the entrance of the Suez Canal port (**Figure 2**).

3.2 Image classification

Unsupervised classification was carried out on the three data sets of the images separately using a histogram peak cluster technique to identify dense areas or frequently occurring pixels [13–15]. Generally, multispectral classification consists of a compression of all information in a multispectral data set into a single image that depicts the major types of surfaces in different colors [16]. Maximum likelihood of supervised classification was applied to detect land cover classes. Once

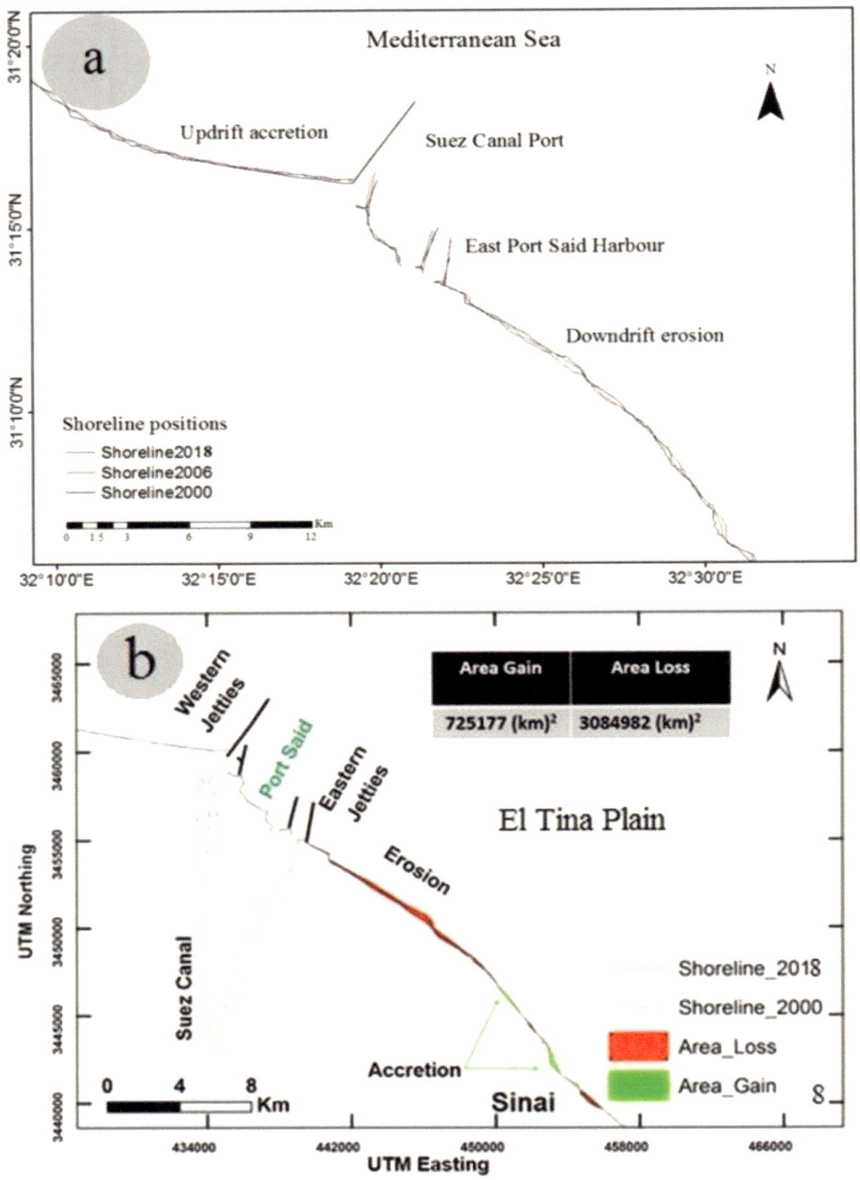

Figure 2.
(a) Patterns of shoreline changes during 2000–2018 and (b) areas of loss and gain along El-Tina plain.

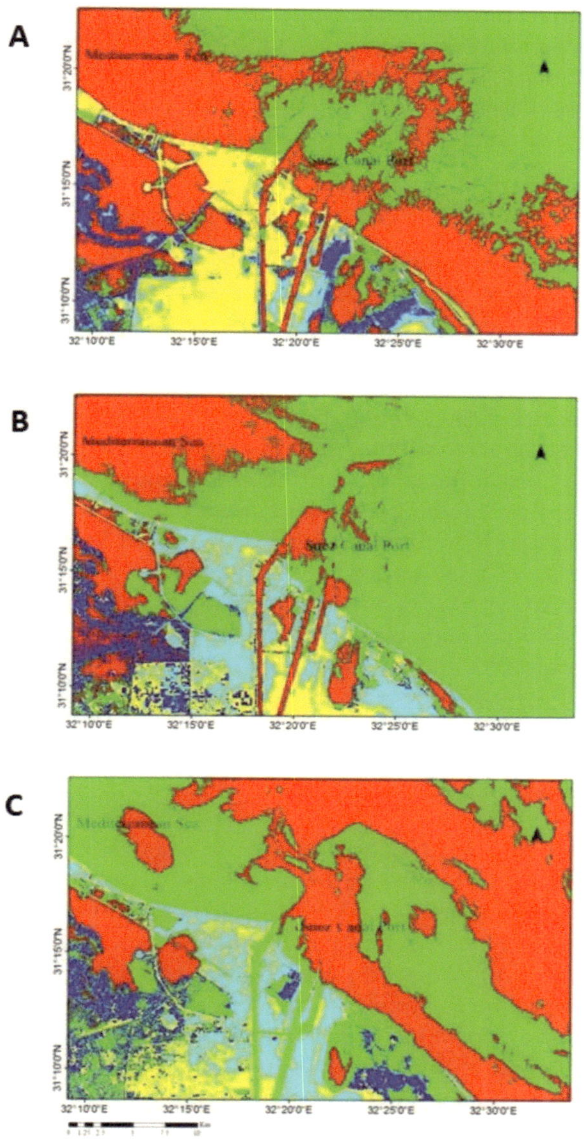

Figure 3.
Image classification during 2000–2018; green color showing siltation along the study site coastal zone.

a sufficient number of such spectral subclasses were acquired for all information classes, a maximum likelihood classification was performed with the full set of refined spectral classes [17]. Image classification of the Enhanced Landsat Thematic Mapper displays an increase in siltation problem along the entrance of the Suez Canal port during 2000–2018 (**Figure 3**).

4. Required data for modeling

The data input for Mike 21-2D modeling, which is the key parameter to run spectral wave (SW) model, the flow model of hydrodynamics (HD), and flow model of sediment transport (ST), includes bathymetry, tide, wind, waves, sediment grain size, and shoreline position. Simulation of shoreline changes, waves, sediment transport, and hydrodynamic regime, using the Mike 21 HD (Flow Model and Hydrodynamic Module), needs some data sets, which are not changed during all simulation analyses. The required data are (4.1) extracted shoreline positions, (4.2) offshore wave parameters, (4.3) bathymetric survey, and (4.4) sediment grain size.

4.1 Extracted shoreline positions

The shoreline positions provided for modeling were extracted from remote sensing results using Thematic Mapper image technique. Wave characteristics required for modeling are wave height, period, and direction. Wave data were measured at Ras El-Bar station. Waves were recorded using a Cassette Acquisition System (CAS); the wave gauge was installed about 1200 m away from the western side of the navigation channel of the Damietta harbor, at water depth of 12 m. The recorder measured the wave characteristics for 20 min each 4 h during a day [18]. Data provided from the Coastal Research Institute in Alexandria have been analyzed in order to determine wave height, period, and direction. These data represent the wave parameters in year 1986 for eastern Nile Delta coast. Bathymetric data were supplied from the Suez Canal Authority. It shows parallel offshore contours to the shoreline trend from 0 to 20 m depth within the nearshore zone.

Satellite images are the main source of data for shoreline positions in this study. Data acquired include SPOT-4 images for year 2006, ETM+ Landsat 7 images during 2000–2012, EgyptSat images for year 2010, and Landsat 8 images for years 2013, 2014, and 2018.

4.2 Offshore wave parameters

Tidal data along the Egyptian Mediterranean coast do not exceed 44–50 cm range. Consequently, tide has insignificant role as input data for MIKE 21 modeling. One-year measurements (1990–2000) of wind and wave series data used in this study were measured in Port Said. The strongest wind series are coming from SSW to WSW direction and blowing from land; therefore, it did not create any waves approaching the shoreline. However, it transported beach sand toward offshore. The velocity of this series is 13.8 m/s. This speed is not strong enough to generate storm (wind speed 18 between 24.5 and 32.6 m/s). The main input wave parameters for the hydraulic computations in LITDRIFT and LITLINE are wave height, wave angle, and wave period. Longshore currents crossing beach profiles are generated using these programs due to shoaling and refraction of the incident waves. Wave data were supplied from many sources such as Suez Canal Authority and Delft Hydraulics. For year 2003, it was measured in Damietta promontory, while wave data for years 1986, 1987, and 1990 were measured in Rosetta promontory. Finally, during 2009–2013, it was measured in Alexandria.

4.3 Bathymetric survey

The bathymetric data used in this study were supplied from the Egyptian Military Survey as hard copy maps. Bathymetric data for year 2004 was used in Port Said and Suez Canal areas. This data was scanned by AutoCAD 2014 software to be digitized and processed using civil 3D software to get (x, y, z) format and work with Mike 21 Flow Model. Mesh file map was generated from the x, y, z digital file using MIKE 21 to understand hydrodynamics regime and sediment transport.

4.4 Sediment grain size

The sediment properties should be defined for each grid point in the cross-shore profile. The average grain size diameter at one of the concerned site, Port Said, is 0.14–1.21 mm (fine sand), and the closure depth is at a range of 2–4 m, and berm height varies from 0.5 to 1 m [18]. The changes in this range produce slight response in shoreline changes (which calculated by Genesis 1D modeling) [19]. When the median grain size decreased from 0.40 to 0.14 mm, there was no change in the shoreline position. The LITPACK module calculates the sediment transport capacity (i.e., it assumes that there is an unlimited source of sediment supply) [20].

5. Results of modeling techniques

Coastal zones are one of the most important areas for human activities and infrastructure growth. However, the systems in these areas are dynamic and need to be studied extensively before planning infrastructure to avoid damages. Numerical modeling is considered as important tool to evaluate coastal zone systems and predict its environmental characteristics. Quantitative prediction of coastal processes and coastal evolution via numerical modeling is now possible due to the major advances that have been made in understanding physical processes and mathematical modeling techniques. The problem of Nile Delta localities is the intensive erosion following construction of some engineering protection and transporting of these materials from beach face by waves and longshore currents. Consequently, the application of modeling is very important to understand hydrodynamic regime and coastal processes controlling coastal erosion and accretion at the concerned sites. In addition, the impact of construction of some engineering protections on the coastal morphodynamic during 2000–2015 will be evaluated. MIKE 21 by Danish Hydraulic Institute (DHI) software is such an integrated complete coastal modeling suite, commercially marketed by Danish Hydraulic Institute, which delivers superior technology, expert support, and outstanding value based on 40 years of experience. The DHI group helped us in this study by giving a permission to use the original package of MIKE 21 with a limited license. Certain modules were selected from MIKE 21 to achieve the objectives of this study; they are (1) MIKE 21 SW, (2) MIKE 21 HD, and (3) MIKE 21 sand transport (ST).

5.1 MIKE 21 spectral wave

Wave characteristics required for modeling are wave height, period, and direction. Wave data were measured at Ras El-Bar station for the eastern part of the Mediterranean coastal zone. The station was put nearly 1200 m away from the western side of the navigation channel of the Damietta harbor, at water depth of 12 m [7, 19]. The recorder measured the wave characteristics for 20 min each 4 h during a day. Data provided from the Coastal Research Institute in Alexandria have been analyzed in order to determine wave height, period, and direction. These data represent the wave parameters in year 1992 for eastern Nile Delta coast.

In order to simulate the growth, transformation of wind-generated waves, and swell in offshore in coastal zones, MIKE 21 spectral analysis module (SW) has been used to get two-dimensional wave heights for the study area according to wave bottom interactions and wave structure interactions (shoaling, refraction, diffraction, reflection, bottom friction, and wave breaking) [20] (**Figure 4**).

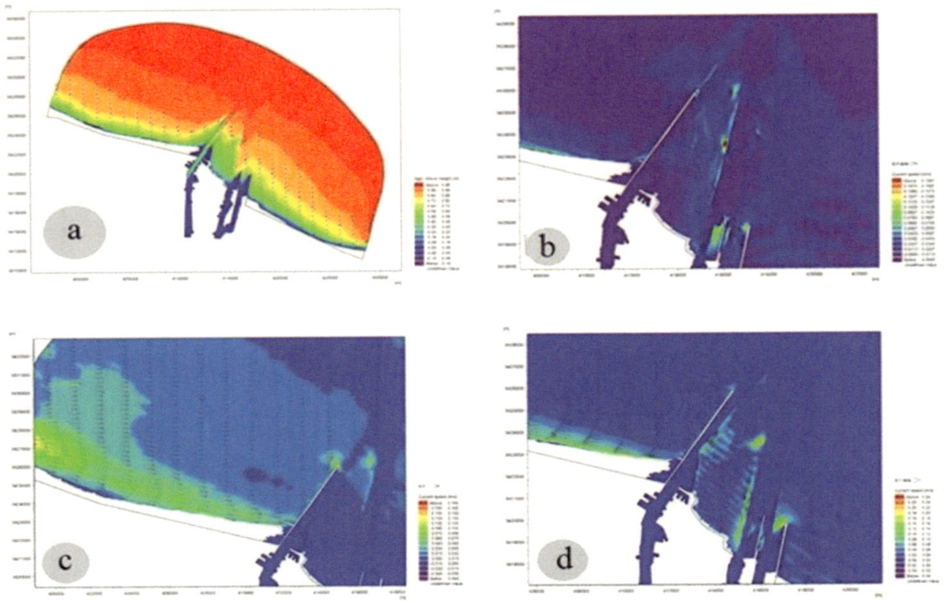

Figure 4.
Simulated spectral wave (SW) model for Suez Canal entrance including (a) spectral waves in NNW direction, (b) spectral waves in NW direction, (c) spectral waves in N direction, and (d) spectral waves in NE direction.

The model has coarsely triangle mesh at offshore zones and finely triangle mesh at surf zones and study area to get more accurate wave heights with acceptable model run period [21].

MIKE 21 SW includes two different formulations:

1. Directional decoupled parametric formulation

2. Fully spectral formulation [22].

5.2 MIKE 21 hydrodynamics

MIKE 21 hydrodynamic has modeled to solve currents due to interaction between wave radiation stresses and water level variations with bottom depths and structures at study area in addition to updrift and downdrift zones. The hydrodynamic forces due to wave breaking are the main effective parameters that lead sediments to move [23].

5.3 MIKE 21 sand transport

This model will be used to predict coastal sand transport and morphodynamics; MIKE 21 sediment transport is designed for the assessment of the sediment transport rates and related initial rates of bed level changes of non-cohesive sediment (sand) due to currents or combined wave-current flow [24]. It is only adapted for non-cohesive sediment (e.g., sand) for which it provides good results. Mathematical shoreline models are tools which are widely used to study the effect of hydrographic parameters on coastal processes and calculate the sediment transport rates and consequently the shoreline changes. The sediment transport process at onshore/offshore and/or alongshore is very complicated problem because it results from

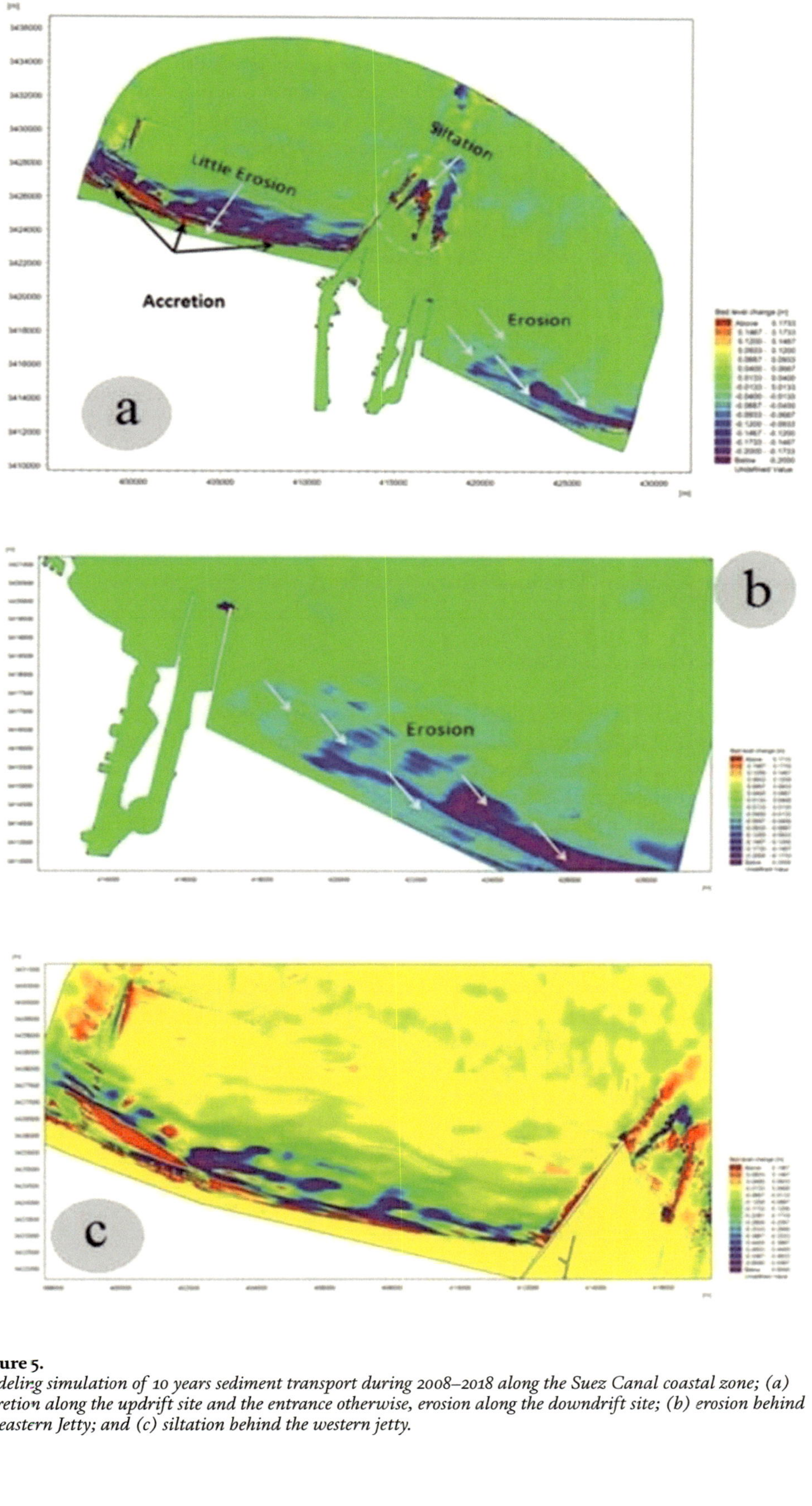

Figure 5.
Modeling simulation of 10 years sediment transport during 2008–2018 along the Suez Canal coastal zone; (a) accretion along the updrift site and the entrance otherwise, erosion along the downdrift site; (b) erosion behind the eastern Jetty; and (c) siltation behind the western jetty.

iteration from wind, waves, and currents with the bottom sediments and/or the shore face. The orbital velocity of the waves is the principal force to shake the sediments in its place and put them in suspended case, while the currents existing in this area are responsible for transport of sediment from one place to another.

Therefore, the sand movement in the longshore direction is the longshore sediment transport, while the actual volume of sand involved in the transport are termed the littoral drift Qs that counted in m^3/year or month. All morphological changes happen due to the littoral drift current, which was created as a result of waves that approach the coastline with an oblique angle. Based on that, the relationship between the incident wave and shoreline orientation is a major factor in evaluating the morphological changes for any studied area. All morphological changes happen due to the littoral drift current, which was created as a result of waves that approaches the coastline with an oblique angle. Based on that, the relationship between the incident wave and shoreline orientation is the goal of this study.

In conclusion, siltation inside Suez Canal entrance can be explained due to moving of waves and currents at west Port Said and inside the entrance of Suez Canal. Therefore, when current comes from shallow depths at west of Port Said in the direction of W-E to deep depths at the entrance of Suez Canal, while it is carrying sediment load, its speed gradually decreases near the long western Suez Canal jetty and starts making eddies up to the long jetty, and once it becomes quiet, it starts throwing its sediment load inside Suez Canal entrance (**Figure 5a–c**).

6. Discussion and conclusion

Construction of ports such as in the Suez Canal entrance has a significant potential effect on natural sediment transport processes. This causes disruption to the adjacent beaches. When current transfers from low contour level at west of Suez Canal long jetty, it decreases gradually inside the entrance of Suez Canal then started to increase and decrease back and forth, by making eddies. This eddies around the eastern and western jetties start to throw their load, while the high current speed at the eastern side of Port Said causes erosion. Consequently, some recommendations are suggested as follows:

- Increasing coastal development has led to a conflict between man desire and nature processes that modified the used land. Therefore, most countries that are located on coastal areas should study in details the coastal zone management problems and risk as a result of the protection work structure effect on coastal area hydrodynamic regime.

- Choice should be taken between allowing unlimited construction of high-valued property and implementing the regulations that prevent developments which would be exposed to major hazards. This choice requires the estimation and prediction of probable future shoreline position and risk assessment to balance between the possible losses of development against the reduction of the existing shoreline.

- Integration between remote sensing and 2D finite hydrodynamic flow models is mandatory to evaluate, interpret, and analyze the effect of costal processes and protection hard structures.

- Because of high current speed at Suez Canal east jetties, the downdrift area is exposed to sever erosion, and there is an investment plan to implement big national projects on it, so it is mandatory to study and monitor the rates of erosion and consider it in any construction plans.

- Studying of sediment transport and bed level change and highlighting the risky and hot spot areas of erosions of east and west of Suez Canal to predict the annual rates before implementing any projects.

Author details

Mona Fouad Kaiser[1]*, Walaa Awaad Ali[2] and Maysara Khairy El Tahan[3]

1 Geology Department, Faculty of Science, Suez Canal University, Ismailia, Egypt

2 Petroloum Geology Department, Faculty of Petroleum and Mining Sciences, Matrouh University, Marsa Matrouh, Egypt

3 Transportation Department, Faculty of Engineering, Alexandria University, Egypt

*Address all correspondence to: monakaiser2013@gmail.com

References

[1] El Tokhi M, Abdel Gawad E, Lotfy MM. Impact of heavy metals and petroleum hydrocarbons contamination of the East Port Said port area, Egypt. Journal of Applied Sciences Research. 2008;**4**(12):1788-1798

[2] Abu Asi IM. The Geographic Criteria for the East Port Said along the Mediterranean. Egyptian Geographical Association; 1998. 29 pp (in Arabic)

[3] Dewidar KM, Frihy OE. Thematic mapper analysis to identify geomorphologic and sediment texture of El-Tineh plain, north-western coast of Sinai, Egypt. International Journal of Remote Sensing. 2003;**24**:2377-2385

[4] Coleman JM, Robert HH, Murray SP, Salama M. Morphology and dynamic sedimentology of the eastern Nile delta shelf. Marine Geology. 1981;**42**:301-312

[5] Frihy OE, Debes EA, El Sayed WR. Processes reshaping the Nile delta promontories of Egypt: Pre- and post-protection. Geomorphology. 2003;**53**:263-279

[6] Kaiser MFM. Monitoring and modelling the impact of engineering structures on coastline change, Nile Delta, Egypt [Ph.D. thesis]. UK: University of Reading; 2004. 270 p

[7] Ali W, Kaiser MF, Kholief S, El-Tahan M. Assessment of shoreline stability and solidity for Please provide volume number and page range for Refs. [7, 12].future investment plans at Ras El-Bar Resort. Egyptian Journal of Aquatic Biology and Fisheries. 2017

[8] Wilson PA. Rule-based classification of water in Landsat MSS images using the variance filter. Photogrammetric Engineering and Remote Sensing. 1997;**63**:485-491

[9] Sonka M, Hlavac V, Boyle R. Image Processing, Analysis and Machine Vision. London: Chapman & Hall; 1993. 555 p

[10] Janssen LLF, Molenaar M. Terrain objects, their dynamics and their monitoring by the integration of GIS and remote sensing. IEEE Transaction on Geoscience and Remote Sensing. 1995;**33**:749-758

[11] Frouin R, Schwindling M, Deschamps PY. Spectral reflectance of sea foam in the visible and near-infrared. In situ measurements and remote sensing implications. Journal of Geophysical Research. 1996;**101**:14361-14371

[12] Ali W, Kaiser MF, Kholief S, El-Tahan M. Assessment of coastal change along Baltim Resort from (2000-2015) using remote sensing and DSAS method. Egyptian Journal of Aquatic Biology and Fisheries. 2017

[13] Eastman JR. Supervised Classification in IDRISI for Windows Version 2, Tutorial Exercises. Worcester, Massachusetts: Clark University; 1997. pp. 86-94

[14] Lillesand TM, Kiefer RW. Remote Sensing and Image Interpretation. 4th Edition. New York: John Wiley & Sons, Inc; 1994. 750 pp

[15] Mather PM. Computer Processing of Remotely-sensing Images. An Introduction. 2nd ed. Chichester: John Wiley and Sons; 1999. pp. 1-75

[16] Lillesand TM, Kiefer RW, Chipman JW. Remote Sensing and Image Interpretation. 5th Edition. New York: John Wiley & Sons, Inc; 2004. 763 pp

[17] Kaiser MF. Monitoring and Modelling the Impact of Engineering

Structures on Coastline Change, Nile Delta, Egypt. Reading, U.K.: University of Reading, Ph.D; 2004

[18] Holthuijsen LH. Waves in Oceanic and Coastal Waters. Cambridge Univesity Press; ISBN 978-0521860284. 2007. pp. 387

[19] Ali W, Kaiser MF, El-Tahan M. Assessment of bottom erosion in front of Rosetta Eastern and Western Groins System. In: The Scientific Committee (SC) of the 1st Scientific Congress of Junior Geosciences in Egypt (SCJGE-1) held at Suhag University, February 3-4-2019. Paper Work in Submission. 2019

[20] NaefD, RickenmannD, RutschmannP, McArdell BW. Comaparison of flow resistance relations for debris flows using a one-dimensional finite element simulation model. Natural Hazards and Earth System Science. 2006;**6**(1):155-165

[21] Julien PY. Erosion and Sedimentation. 2nd ed. xviii + 371 pp. Cambridge University Press; 2010;**148**(04):683-684. DOI: 10.1017/S0016756811000215

[22] El-Tahan M, El Sharnouby B. Dramatic Erosion of Nile Delta Coast Caused by Anthropogenic and natural Influences. Delft. 2015

[23] Wang JD. Numerical modelling of bay circulation. In: The Sea. Ocean Engineering Science. Australia: Elsevier Publisher; Vol. 9. Part B. Chapter 32. 1990. pp. 1033-1067

[24] Eurosion. Living with coastal erosion in Europe: Sediment and space for sustainability. A guide to coastal erosion management practices in Europe: Lessons learned. Coastal erosion–evaluation of the need for action. Directorate general environment. In: European Commission. Prepared by the National Institute of Coastal and Marine Management of the Netherlands. 2004

Chapter 6

Ionospheric Monitoring and Modeling Applicable to Coastal and Marine Environments

Ljiljana R. Cander and Bruno Zolesi

Abstract

Ionospheric monitoring and modeling in costal and marine environment is reviewed and characterized in terms of state of art, global, regional, and local issues across different domains of solar-terrestrial conditions for practical applications. Their effects on critical technological systems are either controlled by the Earth's ionosphere, as in telecommunications and information systems, or simply influenced by its variability, as in trans-ionospheric radio communication, and navigation systems. The evolution of long-distance high-frequency (HF) communications and then still the actuality of HF radio links especially for the coast environment, maritime services, and aeronautical applications, for control and emergency services, for communications equally important in case of great islands and remote areas, for economic reasoning and easy management, and for efficient backup in case of cyber threats are discussed. Some preferred methods for a proper assessment of HF networks have been identified, and examples of existing long-term prediction and near real-time nowcasting in ionospheric space weather modeling to be used, particularly in the Mediterranean area, are presented along with contemporary references.

Keywords: ionosphere, space weather, model, HF and GNSS systems

1. Introduction

Variability in the Earth's ionosphere reduces the reliability of radio-frequency (RF) and global navigation satellite system (GNSS) communication systems because they depend on the attenuation, absorption, reflection, and refraction and accordingly changes in the propagation, phase, and amplitude characteristics of radio waves, in addition to the scintillation phenomenon induced by abrupt variations in electron density along the radio path. Significant scientific work over many decades, within national and international projects, is being conducted on monitoring, proper understanding, and predicting ionospheric variability in order to enhance reliability and robustness of both ground- and space-based communications networks and other applications for the benefit of society [1].

Ionospheric bottom- and topside observations and studies related to fundamental as well as radio communication and navigation purposes cover most of the planet but in an inhomogeneous way. Accordingly, the discovery and complete characterization phase for most ionospheric processes is still in progress. This is particularly

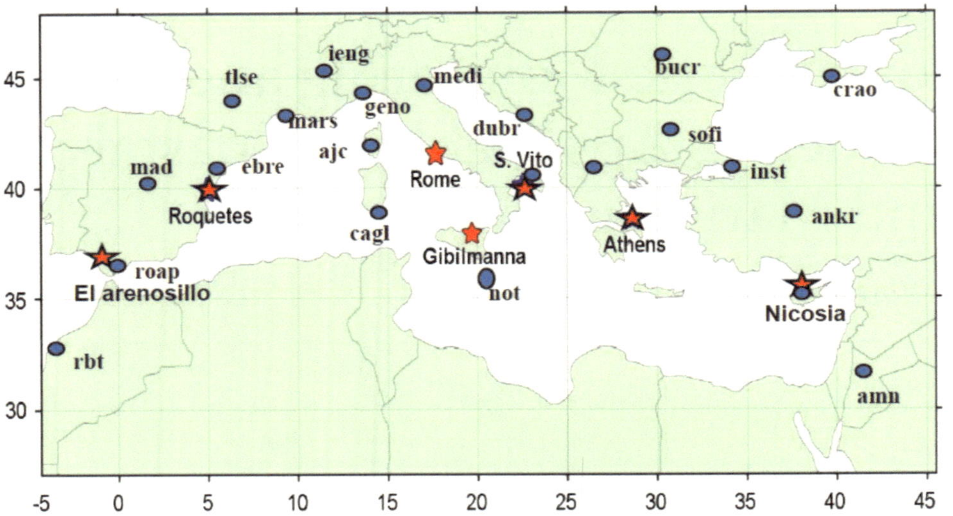

Figure 1.
Map of Mediterranean and north African regions depicting the real-time operating ionosondes (red stars) and GNSS sites (blue points). (www.igs.org/network).

true for the transition area between mid-latitude and equatorial ionosphere where the Mediterranean and North African regions have a special importance for ionospheric studies and applications. Moreover, the sea and deserted areas in these regions make even more important ionospheric monitoring and modeling because of limited availability of sufficient and high-quality data for activities in a broad range of areas within geophysics. **Figure 1** shows the positioning of the ionosondes and GNSS receivers, as principal sources of ionospheric information within Mediterranean and North African regions, respectively.

In order to illustrate ionospheric monitoring and modeling results applicable to coastal and marine environments, Section 2 discusses in brief the salient points of the well-established physical background of the Earth's upper atmosphere. Section 3 describes the basic principle of the main techniques systematically used for monitoring the ionized layers of the Earth's upper atmosphere based on propagation effects that influence radio waves traveling through the ionosphere. Section 4 contains an overview of the current state of the most important electron density models, while particular attention is given to ionospheric mapping techniques to spatially interpolate derived parameters between sites from the sparse network of measurements and/or observations with emphasis on local and/or restricted area. Some aspects of HF communications in coastal and maritime applications are described in Section 5. Finally, Section 6 briefly summarizes the work, notes limitations of the current methodology, and suggests areas for further study.

2. General description of the Earth's ionosphere

The ionosphere is embedded in the neutral Earth's atmosphere beginning at an altitude of about 50 km and extending outward up to 1000 km. It is dynamic plasma medium, highly variable in space on scales of meters to hundreds of kilometers and time on scales of seconds to hour, months, and solar cycles that exhibit climatology and weather features at all latitudes, longitudes, and altitudes. The Earth's ionosphere is created and maintained on a very regular basis by energetic solar irradiance in the extreme ultraviolet (EUV) and X-ray regions of the spectrum that ionizes a part of the neutral atmosphere. Absorption of EUV radiation at other wavelengths also heats a small fraction of the neutral atmosphere so that the deposition of this energy drives a complex cycle of photochemical response that interacts strongly with atmospheric transport. Solar variation of its spectrum on timescales as long as the 11-year solar activity cycle can have a significant effect on ionospheric structure and dynamics, and hence on propagation parameters, in terms of solar cycle, annual, seasonal, daily, and hourly variations [1, 2].

The maximum expansion of the ground-based ionospheric measurements was achieved during the International Geophysical Year (IGY, July 1957 to December 1958) and has steadily continued to the present days. Since 1995 the US Global Positioning System (GPS) has made possible the electron content observations along a radio signal path between a satellite and a ground receiver station, with valuable total electron content (*TEC*) data coming from sustained growth of GNSS technologies. Measured quantities like critical frequencies f_0E, f_0F1, and f_0F2 are related to ionospheric layers, the F2 layer (atomic oxygen ions) around 350 km altitude, the daytime F1 layer (molecular oxygen ions) around 190 km and E layer at 120 km, and the D layer near 70 km (**Figure 2**). Separate regions in the Earth's ionosphere including topside part above 1000 km are direct consequence of solar spectrum energy deposited at various heights depending on absorption of atmosphere, of recombination processes depending on density of atmosphere changeable with height, and of the upper atmosphere composition itself also variable with height. The various forms of temporal and spatial variability of each ionospheric layer include both systematic diurnal, seasonal, and solar cycle variations and large irregular variations. They are also a function of geomagnetic latitude as the role of the Earth's magnetic field is essential, and in solar-terrestrial physics, it is very often described by geomagnetic indices such as *Dst*, *AE*, *Kp*, and *Ap* [3].

From **Figure 2** it is clear that the F2 layer has the greatest plasma density, with maximum electron density $NmF2 = 1.24\ 10^{10} foF2^2$ (see also Eq. 3), which carries the highest frequencies for less absorption, and radio waves can travel the furthest distance with a minimum of attenuating hops making this ionospheric region the most important for HF communications. Its rapid changes throughout the day are shown in **Figure 3** by an example of the day-to-day *foF2* variability at the Nicosia (35.1° N, 33.3° E) ionosonde station over temporal scales from 15 minutes to 1 month

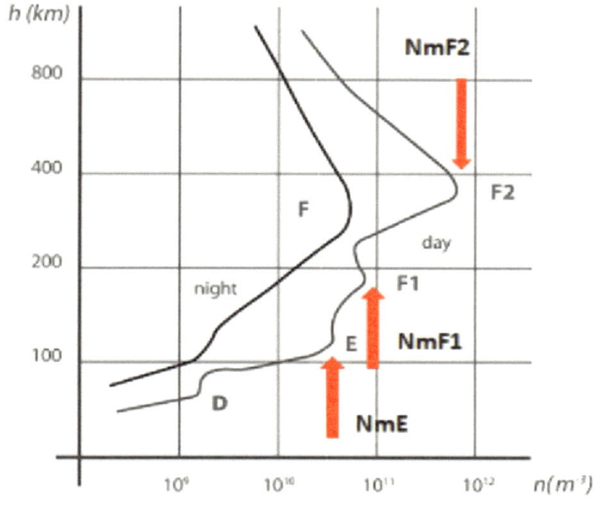

Figure 2.
Diurnal and nocturnal ionospheric N(h) profile representing electron density as a function of height.

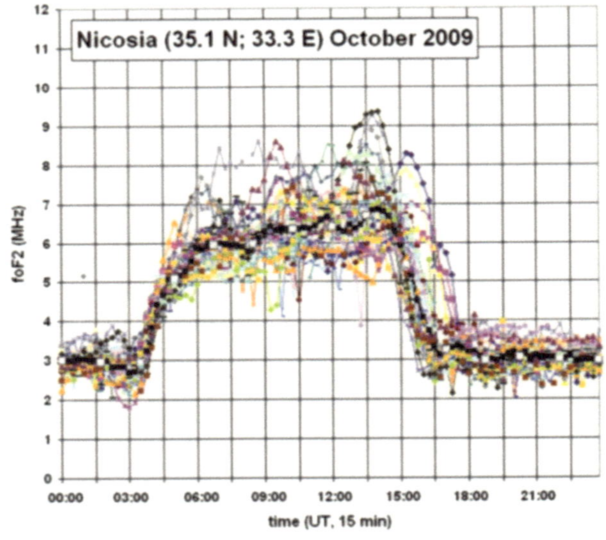

Figure 3.
Plot of the day-to-day critical frequency foF2 *variability over temporal scales from 15 minutes to 1 month at the Nicosia ionosonde station during a period of very low solar activity in October 2009 with the monthly mean of the daily sunspot number Ri = 4.6. The* foF2 *monthly median is in black.*

of October 2009, representing ionospheric equinox during very low solar activity conditions. During this particular month, overall *foF2* departure from the monthly median values is about ±30%, while during very high solar activity conditions, departure from median conditions is usually much greater following the ratio of the corresponding values of solar radio flux *F10.7*, an index derived from measurements of total emission originating from high in the solar chromosphere and lower corona thus frequently used as a very good indicator of solar activity [2].

Similar results of the high variability are obtained if the maximum ionization density is replaced by the column density or vertical total electron content, *VTEC*, of the ionosphere at the co-located GNSS station at the site nico (35.1 N, 33.4 E), in Cyprus (http://www.igs.org/). **Figure 4** shows diurnal 10 minutes *VTEC* values during absolute solar minimum in December 2008 when the variability around the monthly median values is around ±40% during the nighttime and a little lower during the daytime ≈ ± 30%. Again it has to be emphasized that solar cycle dependence is fundamental when values of *VTEC* around solar maximum largely exceed by a factor of approximately 2 for those around solar minimum.

Solar events such as flares and coronal mass ejections often produce large variations in the corpuscular and electromagnetic radiations leading to disturbances of the regular regions known as ionospheric storms. They have important terrestrial consequences generating large disturbances in ionospheric electron density distribution *N(h)*, total electron content *TEC*, and the electric currents system. All these phenomena can continue for a few hours to several days and lead to significant changes in the ionospheric plasma parameters which can be particularly damaging to both satellite- and ground-based systems. The F region's response to ionospheric storms has been studied since the earliest days of solar-terrestrial physics more than 90 years ago. In general results show consistency in characteristic patterns of an ionospheric storm: (1) a short positive phase that occurs during the daytime hours on the first day of a storm with the tendency to significantly increase electron density during the first 24 hours of the storm above its quiet time reference level and (2) a prolonged negative phase on subsequent days leaning to significantly

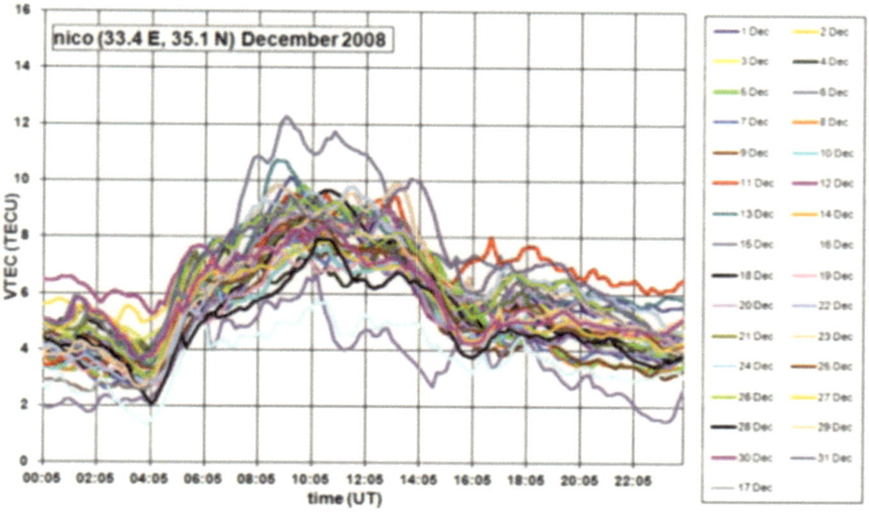

Figure 4.
Plot of the day-to-day vertical total electron content VTEC *variability at the nico GNSS station over temporal scales from 10 minutes to 1 month during absolute solar minimum in December 2008 with the monthly mean of the daily sunspot number Ri = 0.8.*

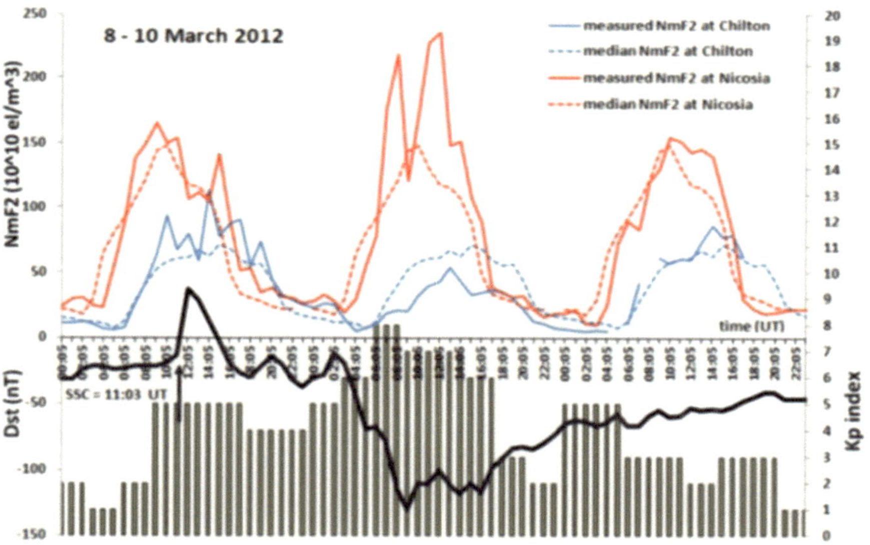

Figure 5.
Time variations in Dst *and* Kp *geomagnetic indices and* NmF2 *at the Chilton (51.6° N, 358.7° E) (blue solid and dashed curves) and at Nicosia (35.1° N, 33.3° E) ionosonde stations (red solid and dashed curves) during the ionospheric storm period of 8 –10 March 2012.*

decrease electron density below its quiet time reference level, with recovery in 1 or 2 days later [see solid and dashed blue curves for the high mid-latitude Chilton (51.6° N, 358.7° E) ionosonde station in **Figure 5**].

Remarkable differences occur in the magnitudes and longevities of the only positive storm pattern represented with solid and dashed red curves for the low mid-latitude Nicosia (35.1° N, 33.3° E) ionosonde station at the island of Cyprus in the Mediterranean Sea. The striking feature is the pronounced *NmF2* increase above monthly median values, which has been taken to represent the quiet reference level at both stations. It is to be believed that short-timescale dynamical mechanisms driving the storms (electrodynamical and thermospheric) dominate the positive phase, while longer-timescale composition changes the negative phase [4, 5].

However a number of questions remain, e.g., solar-terrestrial circumstances and prior storm ionospheric condition necessary for these phases to occur. In particular: (1) duration and magnitude of the negative and/or positive phase versus latitude, local time, season, and phase of solar cycle as well as between different solar cycles and (2) temporal relationships between characteristics of the solar event and the consequent development of the geomagnetic and ionospheric storms in real time. Nowadays they are subjects of intense studies within the space weather domain [2].

3. Ionospheric monitoring

The exploration and the physical description of the ionosphere has been the result of a great activity of experimental observation and continuous systematic monitoring started at beginning of the last century when, G. Marconi realizing on 12 December 1901 a transoceanic radio link, provided the experimental proof of the existence of the Earth's ionosphere postulated during the nineteenth century by various scientists like B. Stewart and A. Schuster. Then the vertical structure of this part of the atmosphere has been described in detail thanks to the technological developments of G. Breit and M. A. Tuve and to the systematic experiments and theoretical studies of Appleton [6].

Two principal methods have been applied to observe and to investigate the terrestrial ionosphere: the first and traditional one is ground-based, the ionospheric vertical sounding by ionosondes to determine electron density of ionospheric plasma as a function of the height, and the second one, more recently, by using geostationary satellites to provide the total electron content.

The first one is a special radar technique based on the principle that when an electromagnetic wave of frequency f penetrates vertically in the ionospheric plasma, the reflection occurs, according to the magneto-ionic theory [7], at the level where the refractive n index becomes zero:

$$n^2 = 1 - (f_N/f)^2 \tag{1}$$

Then considering that the plasma frequency f_N is

$$f_N = \left[\left(Nq^{2)}/(4m\pi^2\varepsilon_0)\right]\right. \tag{2}$$

where N is electron density and q and m are the charge and the mass of the electron, respectively; the reflection in the ionosphere occurs when the incident frequency f is equal to f_N. Furthermore the maximum electron density Nm corresponds to the maximum reflected incidence frequency, called the critical frequency fo:

$$Nm = 1.24\ 10^{10}\ fo^2 \tag{3}$$

where Nm and fo are expressed in el/m^3 and in MHz, respectively.

A vertical ionospheric sounder emits radio impulses with increasing frequency from 1 to 20 MHz, measuring the time delay of radio signals received back from the different ionospheric layer:

$$\Delta t = 2\ h'/c \tag{4}$$

DOI: http://dx.doi.org/10.5772/intechopen.90467

the function of the virtual height of reflection h' is virtual because the signal travels more slowly in the ionosphere than in the free space so that the observed heights h' exceed the true height reflections.

The ionogram, the record produced by the ionosonde, is a plot of the virtual height of reflection vs. the transmitted frequency. In **Figure 6** a typical ionogram is shown produced by a modern digital ionosonde [8], where several important characteristics (like the critical frequencies and the heights of the different ionospheric layers) are indicated as well as the automatic interpretation on the left side. They all have a significant role in the studies concerning ionospheric physics, space weather, and related phenomena.

The routine observations of every ionospheric station need standard techniques and conventions applicable for the interpretation of ionospheric measurements in order to achieve a more phenomenological description of the ionogram as well as provide a simplified description of the ionosphere above the station. They were defined in the URSI handbook of ionogram interpretation and reduction edited by W.R. Piggot and K. Rawer [9]. During the past decades, the ionosondes have had an important technological evolution from the first ones analogical recorded on film, to the digital one, and more recently is the automatic scaling of the ionograms essential for real-time monitoring the ionospheric plasma of space weather purposes [1, 2].

The other principal method of ionospheric observation, the GNSS signals monitoring, is applied to evaluate the ionospheric total electron content, *TEC*, defined as the integral of electron density along the radio wave path s from a satellite transmitter to a ground-based receiver:

$$TEC = \int_s Ne(s)\, ds \tag{5}$$

where Ne is the ionospheric electron density along the path s in electrons/m^3.

This parameter providing information of overall ionization in the ionosphere-plasmasphere system is particularly important for trans-ionospheric communications (propagation at VHF and above), navigation, and solar-terrestrial physics. Considering that the satellite is not at the zenith point of the receiver

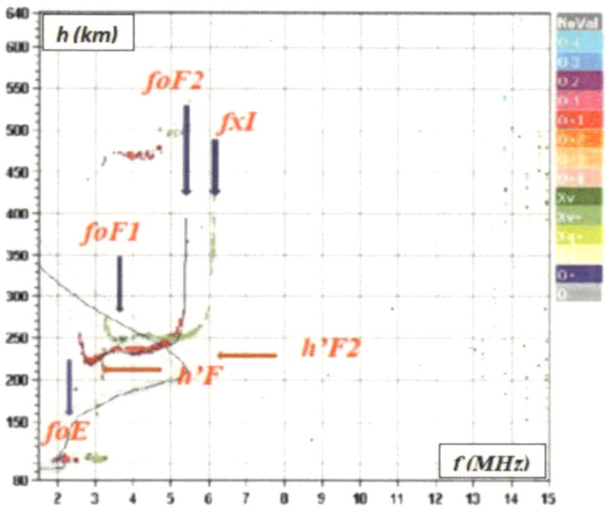

Figure 6.
Daytime ionogram produced by a digisonde, a digital ionosonde, with routinely scaled ionospheric characteristics.

and the real path s is not vertical, it is possible to calculate the vertical *TEC* (*VTEC*) by using different geophysical models to convert the values of the so-called slant *TEC*.

The technique to evaluate the *TEC* is based on the physical fact that the signal propagation time between the satellite- and the ground-based receiver, due to the anisotropic nature of the ionosphere-plasmasphere system, is directly proportional to the total number of free electrons along the signal path. The great increase and technological development on the satellite navigation and positioning system provided a new source of ionospheric data available at several global and regional centers of the International GNSS Service formerly the International GPS Service for Geodynamics (IGS) (http://igscb.jpl.nasa.gov/) in the receiver independent exchange (RINEX) format [10]. These observations are particularly important in the evaluation of the error due to the ionospheric propagation delay for the single GNSS frequency that is inversely proportional to the square of its carrier frequency but proportional to *TEC* along the ray path. The Center for Orbit Determination in Europe (CODE) from Universität Bern (http://aiuws.unibe/ch/ionosphere) regularly provides global *VTEC* maps, while the International GNSS Service makes available an extensive variety of GNSS open data and ionospheric open products (http://www.igs.org/).

4. Ionospheric modeling and mapping

The ground-based and satellite routine measurements constituted, in the second half of last the century, the basis for the global, regional, and local modeling of the terrestrial ionospheric plasma. This activity was supported by international organizations, like the International Union of Radio Science (URSI), the Committee on Space Research (COSPAR), and in particular the International Radio Consultative Committee (CCIR) establishing internationally agreed global propagation models.

Simple models of the lower layers E and F1 are defined as Chapman layers, because referred to an ideal ionosphere as function of the solar zenith angle χ, then the geographical position, and of a solar activity index R [11]:

$$foE\ (\chi, R) = 3.3\ [(1 + 0.0088R)\ \cos\ \chi]^{1/4} \tag{6}$$

$$foF1\ (\chi, R) = 4.25[(1 + 0.015R)\ \cos\ \chi]^{1/4} \tag{7}$$

where *foE* and *foF1* are the critical frequencies in MHz.

However, essential for theoretical studies and practical application are 3D pictures of the terrestrial ionosphere generated by combining the models of the electron density profile, the concentration of the electrons vs. the altitude (see **Figure 2**), and the global and regional mapping of the principal ionospheric characteristics. After the well-known and widely used model introduced by P.A. Bradley and J. Dudney [1], important results were obtained by more general empirical International Reference Ionosphere (IRI) model [12]. Following the beginning of IRI project in 1968, this global model has been systematically improved and updated over time, so that it is currently accepted as the standard for ionospheric parameters in the altitude range from 60 to 2000 km.

Thanks to the first use of computing devices, able to manage the enormous amount of observations collected during the years around the IGY, a numerical method was developed in the Institute of Telecommunication Sciences (ITS) at the Boulder Laboratories of the U.S. Department of Commerce by W.B. Jones and R.M.

Gallet [13] to produce global maps of the two key ionospheric characteristics related to the maximum electron density of the ionospheric F region. They are the median monthly hourly values of *foF2* and *M(3000)F2*, obtained from ionograms of the worldwide ionosonde network. These maps are extremely important for long-distance HF communications representing a significant tool for applied science and for radio users, especially frequency planners at radio broadcasting agencies as well as for geophysicists of the upper atmosphere. Here *M(3000)F2* is the transmission *M* factor (also known as the propagation, obliquity, or maximum usable frequency (*MUF*) factor), an ionospheric characteristic derived from an empirical estimate of the relationship between reflecting layer height, frequency, and oblique radio wave propagation path length [9]. The other important propagation and prediction quantity is the maximum usable frequency, a function of a critical frequency *fo* and an appropriate *M* factor for a given distance *d*:

$$MUF\ (d) = fo\ M\ (d) \tag{8}$$

To produce regional models of the ionosphere for long-term prediction, nowcasting or even short-term forecasting with accuracy much better of the global mode was the target of European projects promoted by the European scientific framework, European Cooperation in Science and Technology (COST) [14]. In **Figure 7** an example of the hourly foF2 nowcast map is given generated by using simplified ionospheric regional model real-time updated (SIRMUP) model for the digital upper atmosphere server (DIAS) [15], an application also embodied in the project ESPAS, the near-Earth space data infrastructure for e-Science (https://www.espas-fp7.eu/).

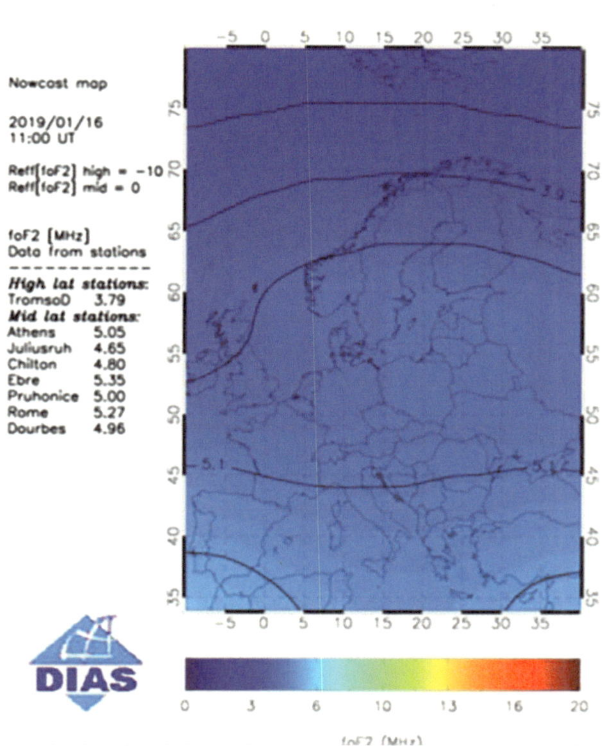

Figure 7.
An example of the foF2 *nowcast map predicted by SIRMUP model in 16 January 2019 at 11:00 UT for DIAS (http://www.iono.noa.gr/Dias/).*

5. HF communications in coastal and maritime applications

HF radio links via ionospere in the 3–30 MHz band represented in most part of the twentieth century the only way for long-distance radio communications, largely used by military and civilian users. Consequently the scientific research in ionospheric radio propagation and monitoring was mainly supported by those countries having global interests and among them the air and maritime communications [16]. The new and great increase of satellite use for long-distance communications gave, between the end of the 1970s and the beginning of the 1980s, the impression that the HF radio communication via ionosphere should be rapidly obsolete. Instead, the use of HF still plays a very important role during emergency situation as the natural catastrophes, for naval or coast to island communications, for people sparse in large extension of country, and for military and civilian radio links located in valleys of a mountain region. So the prediction, forecasting, or even the nowcasting of the future status of the reflectivity of the ionospheric layers is crucial for radio planners to choose the best radio frequency to use or, more recently to know, the evolution of the overall space weather conditions [2].

In a typical HF radio link via ionosphere (**Figure 8**), radio users need to know in advance the range of the useful radio frequencies to be applicable for their service and the area covered by them. The spectrum of the radio frequencies between two points is included between the maximum usable frequency and the lower usable frequency (*LUF*). The *MUF* depends only on the geometry of the radio link and on the conditions of reflectivity of the ionosphere, practically the critical frequency *fo* of the ionospheric layer, while the *LUF* depends, besides geophysical parameters

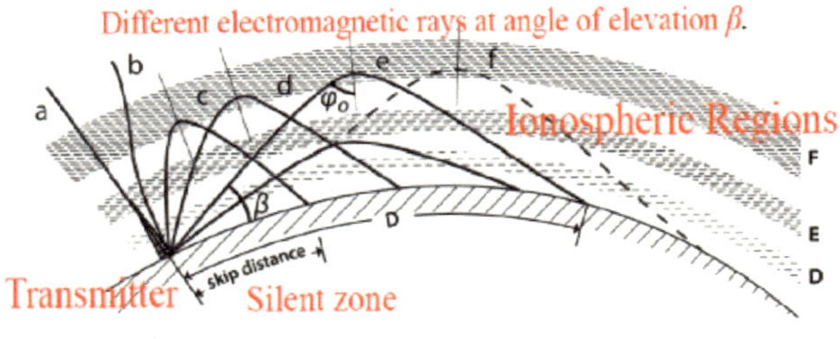

Figure 8.
Simple scheme of an ionospheric radio link.

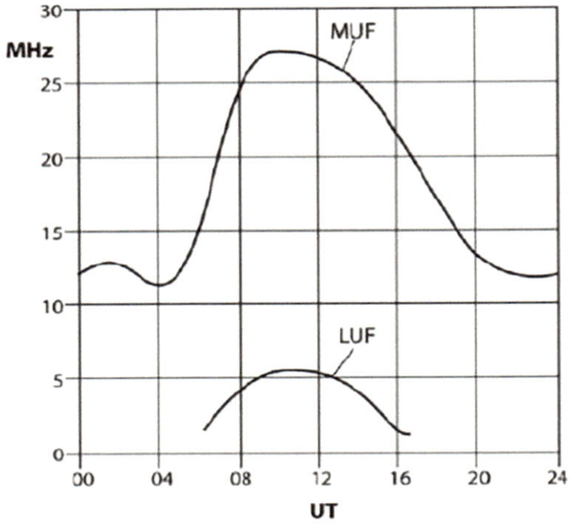

Figure 9.
Example of the HF ionospheric long-term prediction of MUF *and* LUF *for a point-to-point radio link.*

and geometry, principally on the radio technical characteristics of the equipment like power of the transmitter, sensitivity of the receiver, radio noise, gain of the antenna system, etc. [1, 11].

Different national and international institutions provide long-term prediction of the hourly behavior of *MUF* and *LUF* for a given radio link. See as an important example the many radio and space weather information provided by the Australian SWS-Radio and Space Services at https://www.sws.bom.gov.au/, the American NOAA with the IONCAP procedure at ftp.ngdc.noaa.gov/STP/IONOSPHERE/MODELS/IONCAP/, the France Telecom at www.iono.enst-bretagne.fr, and the already mentioned DIAS/ESPAS services. Other national institutions also provide ionospheric prediction, for example, the Italian INGV (Istituto Nazionale di Geofisica e Vulcanologia), inside European organizations or by special request of

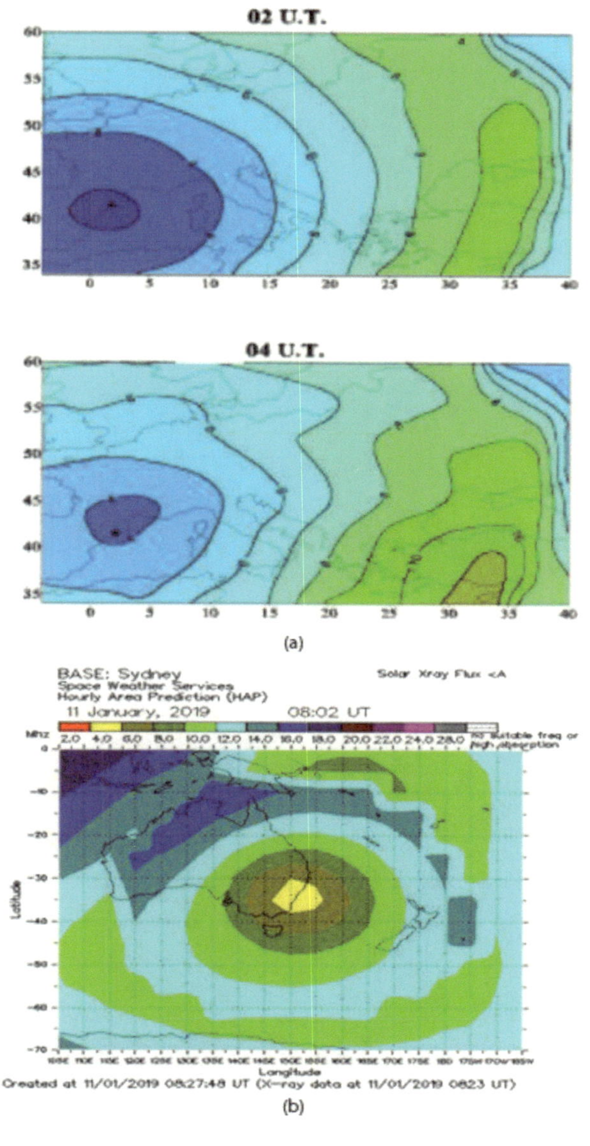

Figure 10.
Skip distance maps or MUF *isolines in two typical examples provided by national services: (a) applied to the Mediterranean area and (b) applied to the Australian region provided by the Australian radio and space services at https://www.sws.born.gov.au › HF_Systems.*

their users. For example, in **Figure 9** a schematic pattern of the *MUF* and *LUF* hourly monthly median predictions for a generic distance and month is shown.

Another parameter extremely relevant to the class of users like the broadcasting agencies and air and maritime application is the skip distance, defined as the minimum distance reflected from the ionosphere, drawn by isolines around the transmitting point. This parameter, typical for radio links from a fixed point to a mobile receiver, is derived by the *MUF* and gives information on the area covered by a given frequency; in fact within this distance, also known as the silent distance, only ground wave propagation is possible.

In **Figure 10** there are two examples of this kind of service. The first one, on the upper panel, applied to the Mediterranean area, gives the hourly isolines of the *MUF* in MHz or the skip distance variable with time for a point of transmission located in South East. This sequence of maps clearly shows the effect of the Sun,

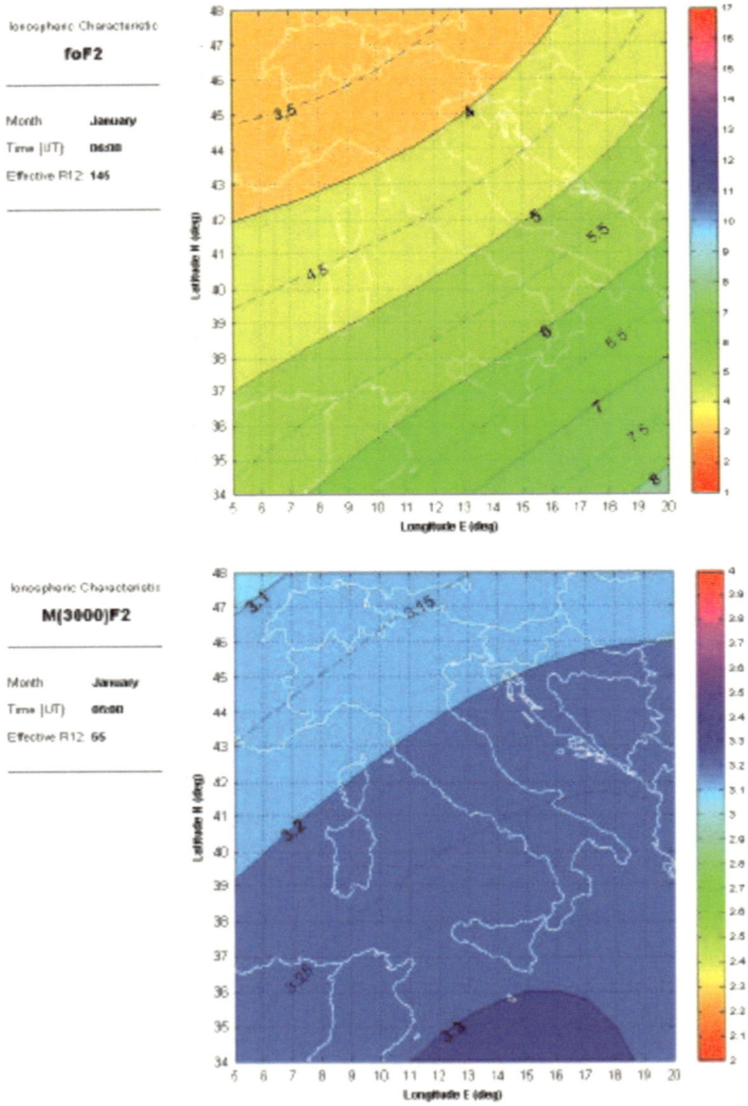

Figure 11.
Examples of foF2 *and* M(3000)F2 *nowcast maps predicted by SIRMUP model for the Central Mediterranean area at two different hours and effective solar activity index* $R_{12}eff$.

increasing the electron density from the South East. The second one, on the lower panel, provided by the Australian Radio and Space Services gives the nowcast optimum recommended frequency for the Australia region and the close-up oceans having a transmitter located at Sydney (http://www.sws.born.gov.au › HF_Systems).

Of course the first and most important application of the ionospheric radio propagation in the coastal and maritime communication is related to the point-to-point radio links between the country and the islands establishing a continued contact between the government critical infrastructures when they are not covered, due to the distance, by other options like VHF radio bridges or ground wave propagation. Secondly, HF radio communication is obviously still important for constant communication of civilian radio users, i.e., the small boats of fishermen and even for the national Coast Guard boats, especially in the Mediterranean area where there are recent operating rescue actions far from their country coast. The Mediterranean area is particularly interesting for the ionospheric physics and radio propagation, not only for historical, economical, and political reasons, but also because in that area, there are the southernmost systematic ionospheric soundings when no other ionospheric observations are available in all the northern part of the African region.

In **Figure 11** two nowcasting maps of *foF2* and *M(3000)F2* are shown produced by the Geomagnetic Indices Forecasting and Ionospheric Nowcasting Tools

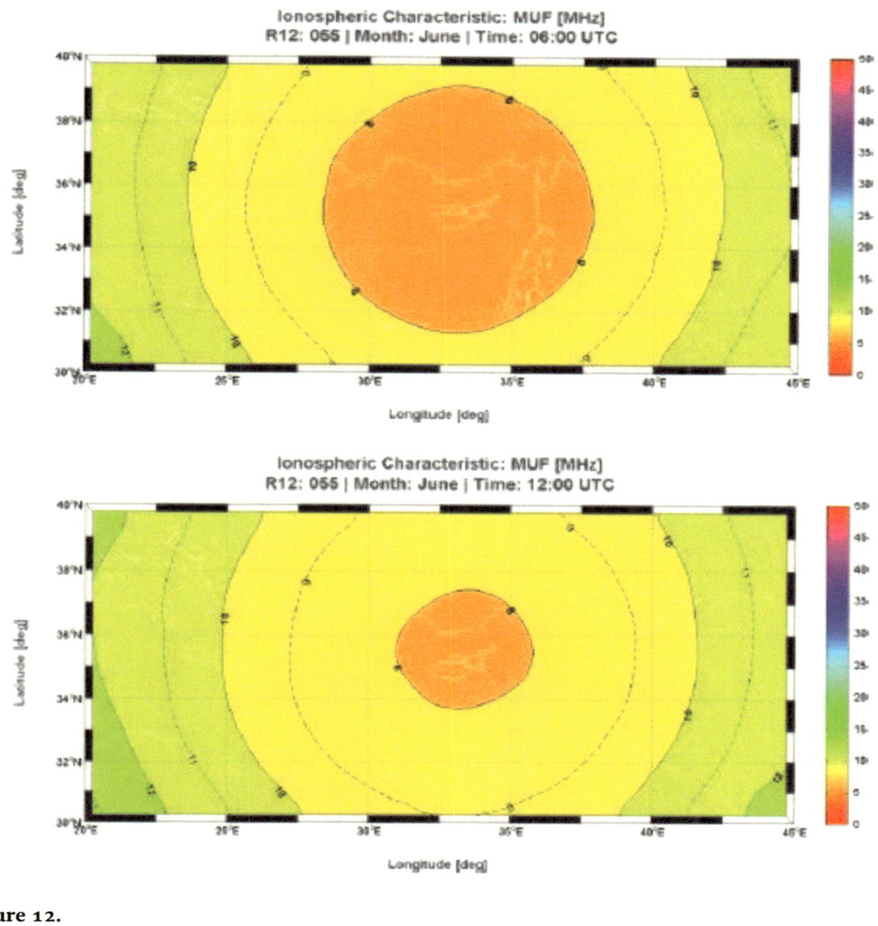

Figure 12.
Example of long-term maps of MUF *over the eastern part of the Mediterranean Sea predicted by SIRM for the Cyprus Ionospheric forecasting service.*

(GIFINT) [17], one of the space weather pilot projects promoted by the European Space Agency in the central area of the Mediterranean Sea.

Long-term maps of the *MUF* in the Eastern part of the Mediterranean area are available within the Cyprus Ionospheric Forecasting Service (CIFS) [18] project promoted by the Frederick Research Center of Nicosia, Cyprus, in **Figure 12**.

The technique involved in the over-the-horizon (OTH) ionospheric radar (**Figure 13**) uses HF frequencies reflected by the ionosphere to detect objects at very long distances, not covered by the ordinary radars that cannot operate beyond the horizon [19, 20]. This technique needs a very high level of energy transmitted (from hundreds of MWatt to GWatt) and a large and complex structure of the antenna system (hundreds of square meters). A real-time control of the ionosphere by a network of ionospheric vertical soundings together with the 3D image and the ray tracing model of the ionosphere is also necessary. **Figure 14** gives an example of 3D image of the ionosphere in the Mediterranean region applicable accordingly [21].

The OTH ionospheric radar, besides the obvious military use, has two important applications from the point of view of the coastal environment. The first one is the control of naval traffic in the space around the territorial waters in other words the border control. The second one is the remote control of the status of the sea level in order to detect tsunami waves for an early alert [22].

Finally, another important application of the HF ionospheric communication has been described within objectives of the European project Short Wave Critical Infrastructure Network based on New Generation (SWING) of high survival radio communications system [23]. The SWING project performed a study to maintain a high survival HF radio network (data/voice) in the real-time support of European critical infrastructure communications. This operating activity should establish a minimum flux of essential information for the management and control, in case, of

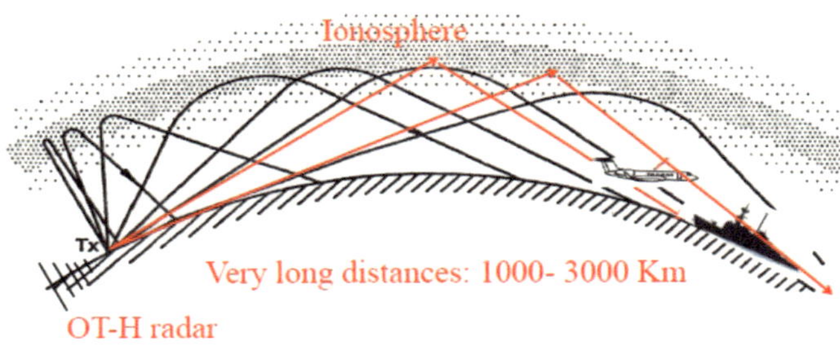

Figure 13.
Simple scheme of the OTH operation for long-range detection of ships, aeroplanes and sea surface conditions.

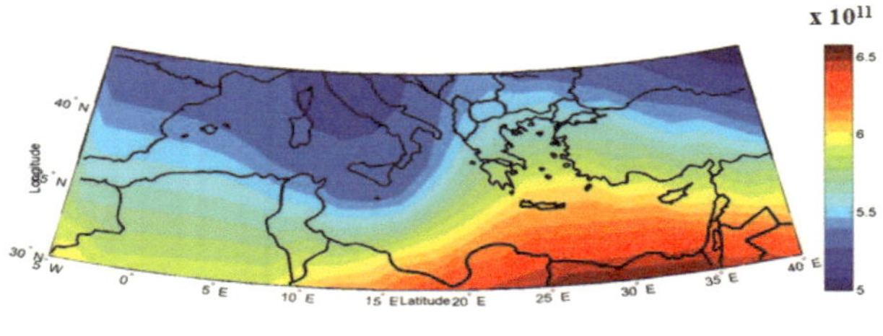

Figure 14.
Map of the ionospheric electron density in el/m^3 at the fixed height of 201 km obtained by the IRI-SIRMUP-P procedure for a given epoch.

wide scale of natural disasters or cyberattacks to render Internet links useless between the great islands of the Mediterranean region and their respective government organization.

6. Conclusions

The land and maritime mobile community for communication of voice and data, in coast stations, ship-ship, shore-ship, and ship-shore modes of operation, occupied about 15% of available HF radio spectrum. High-frequency transmissions and prediction support both maritime safety information (MSI) and distress related communications using digital selective calling (DSC). These communications take place across the maritime mobile service bands within 1.6–26.5 MHz as defined by the International Telecommunication Union (ITU) Radio Regulations [24]. More importantly there is currently considerable worldwide effort being applied to further expand the use of GNSSs by civilian users in general and the civil aviation community in particular. This effort is being directed toward switching from systems under military control to systems under civil control. Scientific studies and technical reports have supported a variety of work in these areas, but here focus is on the Earth's ionosphere's important role. One of the main reasons is related to the fact that this ionosphere is a medium of communication inexhaustible, not polluting and extremely economic especially for its role of possible backup in case of blackout of other systems.

However, the highly complex nature of the Earth's ionosphere, and its potential for huge spatial and temporal variability, is such that a very large number of modeling scenarios is required in general and coastal and marine environments in particular. It has been briefly shown that the algorithms used in this study can be tuned and optimized so as to meet the basic requirements, even under the worst-case space weather conditions [25]. The SIRM, and its real-time updating version SIRMUP, provides regional type of a self-consistent model initialization specifying most important ionospheric characteristics *foF2* and *M(3000)F2* at a given time, while the GIFINT approach could assess specifications and forecasts of ionospheric variables on a local level. This further enhances the necessity for a large number of varied scenarios to be used for verification purposes.

Problems of data shortage, within the Mediterranean and North African regions, and a potential lack in confidence in the performance of models based on such limited data sets have taken on greater importance in recent years. As real-time system operations and integrated management are becoming increasingly present in many domains within geophysics, which requires an increased amount of data with high spatiotemporal resolution, synthetic data is required to augment recorded data and to ensure that a wide variety of ionospheric conditions are tested and an associated model is verified.

Acknowledgements

Data sources are acknowledged as follows: the Space World Data Centre for Solar-Terrestrial Physics (STP) at STFC Rutherford Appleton Laboratory for operation of the ionosonde at Chilton and data access via (http://www.ralspace.stfc.ac.uk/RALSpace/); the Cyprus digital ionosonde station in Nicosia, the Helmholtz Centre Potsdam of GFZ, and the German Research Centre for Geosciences for the production of Kp data (http://www.gfz-potsdam.de/en/kp-index/); the WDC for

Geomagnetism, Kyoto, for the production of Ds index (http://wdc.kugi.kyoto-u.ac.jp/); and the International GNSS Service (IGS) for providing GNSS open data (http://www.igs.org/).

Conflict of interest

None.

Author details

Ljiljana R. Cander[1]* and Bruno Zolesi[2]

1 Rutherford Appleton Laboratory, Harwell Oxford, UK

2 Istituto Nazionale di Geofisica e Vulcanologia, Rome, Italy

*Address all correspondence to: ljiljana.cander@stfc.ac.uk

References

[1] Zolesi B, LjR C. Ionospheric Prediction and Forecasting. Heidelberg, New York, Dordrecht, London: Springer; 2014

[2] LjR C. Ionospheric space weather. In: Springer Geophysics. Cham, Switzerland: Springer Nature Switzerland AG; 2019

[3] Lanza R, Meloni A. The Earth's Magnetism, an Introduction for Geologists. Heidelberg: Springer-Verlaig; 2006

[4] Prölss GW. Physics of the Earth's Space Environment: An Introduction. Berlin, Heidelberg: Springer; 2004

[5] Mendillo M. Storms in the ionosphere: Patterns and processes for total electron content. Reviews of Geophysics. 2006;**44**. DOI: 10.1029/2005RG000193

[6] Appleton EV. Some notes on wireless methods of investigating the electrical structure of the upper atmosphere. I. Proceedings of the Physical Society. 1928;**41**:43-59

[7] Ractliffe JA. The Magneto-Ionic Theory and its Applications to the Ionosphere. Cambridge: Cambridge University Press; 1962

[8] Reinisch BW, Huang X. Automatic calculation of electron density profiles from digital ionograms 3, processing of bottomside ionograms. Radio Science. 1983;**18**:477-492

[9] Piggot WR, Rawer K. U.R.S.I Handbook of Ionogram Interpretation and Reduction. World Data Center a for Solar Terrestrial Physics-Report UAG-23. Asheville: NOAA, Environmental Data Service; 1972

[10] Dow JM, Neilan RE, Rizos C. The international GNSS service in a changing landscape of global navigation satellite systems. Journal of Geodesy. 2009;**83**:191-198. DOI: 10.1007/s00190-008-0300-3

[11] Davies K. Ionospheric Radio. IEE Electromagnetic Waves Series 31. London: Peter Peregrinus Ltd; 1990

[12] Bilitza D. International reference ionosphere 2000. Radio Science. 2001; **36**:261-275

[13] Jones WB, Gallet RM. Ionospheric mapping by numerical methods. Telecommunication Journal. 1960;**12**: 260-264

[14] Zolesi B, LjR C. The role of COST actions in unifying the European ionospheric community in the transition between the two millennia. History of Geo- and Space Sciences. 2018;**9**:65-77. DOI: 10.5194/hgss-9-65-2018

[15] Zolesi B, Belehaki A, Tsagouri I, LjR C. Real-time updating of the simplified Ionospheric regional model for operational applications. Radio Science. 2004. DOI: 10.1029/2003RS002936

[16] Anduaga A. Wireless & Empire. Oxford: Oxford University Press; 2009

[17] Pallocchia G, Bertello I, Amata E, Consolini G, Pezzopane M, Zolesi B, et al. The GIFINT Space Weather products. Conference paper, First European Space Weather Week, Noordwijk; 2004

[18] Pezzopane M, Zolesi B, Pietrella M, Haralambous H, Oikonomou C, Cander LjR. Ionospheric Prediction and Forecasting Services in Mediterranean Area. Conference paper, General Assembly of the European Geophysical Union, Wien, April 2014

[19] Reinisch BW, Haines DM, Bibl K, Galkin I, Huang X, Kitrosser DF, et al.

Ionospheric sounding in support of over-the-horizon radar. Radio Science. 1997;**32**(4):1681-1694

[20] Francis DB, Cervera MA, Frazer G. Performance prediction for design of a network of sky wave over-the-horizon radars. IEEE Aerospace and Electronic Systems Magazine; 2017;**32**(12):18-28. DOI: 10.1109/MAES.2017.170056

[21] Pezzopane M, Pietrella M, Pignatelli A, Zolesi B, Cander LR. Assimilation of autoscaled data and regional and local ionospheric models as input sources for real time 3D international reference ionosphere modeling. Radio Science. 2011;**46**. DOI: 10.1029/2011RS004697

[22] Artru J, Lognonné P, Occhipinti G, Crespon F, Garcia R, Jeason E, et al. Tsunamis detection in the ionosphere. Space Research Today. 2005;**163**:23-27

[23] Zolesi B, Bianchi C, Meloni A, Baskaradas JA, Belehaki A, Altadill D, et al. "SWING": A European project for a new application of an ionospheric network. Radio Science. 2016. DOI: 10.1002/2016RS006037

[24] The Radio Regulations as adopted by the World Radiocommunication Conference (WRC-15, Geneva), Edition of 2016

[25] Lanzerotti LJ. Space weather: Historical and contemporary perspectives. Space Science Reviews. 2017;**212**:1253-1270. DOI: 10.1007/s11214-017-0408-y